Memoirs of the American Mathematical Society

Number 160

Charles W. Neville

Invariant subspaces of Hardy classes on infinitely connected open surfaces

Published by the
AMERICAN MATHEMATICAL SOCIETY
Providence, Rhode Island

VOLUME 2 · ISSUE 1 · NUMBER 160 (last of 2 numbers) · MAY 1975

Abstract

We generalize Beurling's theorem on the shift invariant subspaces of Hardy class H^2 of the unit disk to the Hardy classes of admissible Riemann surfaces. Essentially, an open Riemann surface is admissible if it admits enough bounded multiple valued analytic functions. The class of admissible surfaces contains many infinitely connected surfaces, and all finite surfaces, but does not contain all plane regions admitting sufficiently many bounded analytic functions to separate points. We generalize the theorem of A.H. Read and the Cauchy integral formula to the boundary values, on the Hayashi boundary, of functions in the Hardy classes of admissible surfaces.

A M S (M O S) 1970 subject classifications. Primary 46J15,30A98. Secondary 30A86, 46J20.

Key words and phrases. Abstract boundary, admissible surface, Beurling's theorem, Blaschke region, bounded weak-star topology, Cauchy integral formula, exterior function, Hardy class, Hayashi boundary, inner function, interior function, infinitely connected Riemann surface, invariant subspaces, locally analytic modulus, modules of analytic functions, multiple valued analytic function, periods of differentials, rings of analytic functions, Read's theorem, shift operators.

ISBN 0-8218-1860-0

CONTENTS

Introduction

Beurling's celebrated theorem on the shift invariant subspaces of Hardy class H^2 may be stated as follows [3]:

Theorem: Let H^2 denote Hardy class H^2 for the unit disk. Let S be a norm-closed subspace of H^2 such that $zS \subseteq S$. Then there is a unique inner function I such that $S = I \cdot H^2$.

In this memoir we shall extend Beurling's theorem to the Hardy classes of a wide variety of open surfaces, including many of infinite connectivity. Of course, Beurling's original theorem has been generalized in many directions (c. f. [10], [18], and [22] for surveys). In fact, Sarason in his A. M. S. Memoir obtained a generalization to the Hardy classes of the annulus [40], and Voichick obtained a generalization to the Hardy classes of a finite Riemann surface, that is a compact Riemann surface with finitely many closed, disjoint, conformal disks punched out [45]. Our extension of Beurling's theorem is distinguished by the fact that it applies to a wide variety of open surfaces, including many of infinite connectivity.

To carry out this extension, we need three things: an analogue to Hardy class H^p, a replacement for the inner functions, and a suitable class of open surfaces. First, suppose R is a hyperbolic Riemann surface and let a_0 be a fixed point of R. For $1 \leq p < \infty$, Hardy class $H^p(R)$ is the class of all analytic functions f on R such that $|f|^p$ has a harmonic majorant. Of course, $H^\infty(R)$ is the class of all bounded analytic functions on R. If we norm $H^p(R)$ by setting

Introduction

$$\|f\| = (\text{L. H. M. } |f|^p \Big|_{a_0})^{1/p} ,$$

then $H^p(R)$ is a Banach space under pointwise operations, for $1 \leq p < \infty$. Here, L. H. M. stands for least harmonic majorant. If R is the unit disk and $a_0 = 0$, we obtain the classical Hardy classes and H^p norms. This generalization of the Hardy classes is due independently to Parreau and Rudin [29], [38]. Maurice Heins has recently given an extremely readable account of the $H^p(R)$ spaces [17].

For an analogue to the inner functions, we must either dispense with analyticity or single-valuedness. We have chosen the former course. In chapter 2, we shall define a locally analytic modulus, or l. a. m., to be a real valued function g which is locally the modulus of an analytic function, and an inner l. a. m. to be an l. a. m. g such that $\log g$ is the sum of Green's functions and a singular harmonic function in the sense of Parreau ([29], c. f. also [5, chapter 2] or [17, pp. 6 - 9]).

For a suitable class of open surfaces, we have chosen a class which we shall call, for lack of a better name, the admissible surfaces. The precise definition of an admissible surface will be given in chapter 5, and in chapter 8 we shall show that the class of admissible surfaces contains both planar and non-planar surfaces of infinite connectivity. Essentially, R is admissible if it admits lots of bounded, multiplicative multiple-valued functions, and if the Green's function for R with pole at a_0, $G(a_0,z)$, is sufficiently well behaved.

The generalization of Beurling's theorem, which we shall present in chapter 7, may be stated as follows:

Theorem: Let $1 \leq p < \infty$, and let R be an admissible surface. Let V be a norm closed subspace of $H^p(R)$ such that $gV \subseteq V$ for each $g \in H^\infty(R)$. Then there is a unique bounded inner l. a. m. I such that

$$V = \{f \in H^p(R): \ (|f|/I)^p \text{ admits a harmonic majorant}\}.$$

Conversely, each set V of the above form is a norm closed subspace of $H^p(R)$ such that $gV \subseteq V$ for each $g \in H^\infty(R)$.

In addition, in chapter 7 we shall present a similar generalization of the Rubel-Shields theorem on the β closed ideals of H^∞ [35]. (It will be most convenient to quote the Rubel-Shields theorem after we have discussed the β topology in Chapter 1.) Regrettably, the path to the extended Beurling-Rubel-Shields theorem is long and arduous. However, it is not without points of independent interest. Foremost among these are the developments of chapter 6, where we shall present a version of the Cauchy integral formula applying to the boundary values (on the Hayashi boundary [13]) of a suitable class of analytic differentials, and a converse theorem, which is a generalization of the theorem of A. H. Read [32]. Our proofs of these two theorems carry over, almost without change, to the Martin boundary. However, no knowledge of abstract boundaries is required to read this paper. In chapters 3 and 4, we shall present a self contained exposition of the needed features of the Hayashi boundary using the Gelfand-Neumark theorem and an operational calculus. We shall do this by endowing $h^\infty(R)$, the Banach space of bounded complex valued harmonic functions, with the structure of a B* algebra. The Hayashi boundary will turn out to be the maximal ideal space of $h^\infty(R)$.

We also have a number of negative results and counterexamples, which it will be most convenient to discuss at the end of chapter 1, after we have discussed the Rubel-Shields theorem.

This paper is a generalization of the results established in my doctoral dissertation at the University of Illinois [26]; and announced in [27]. I want to thank my advisor, Lee A. Rubel, for his generous and effective help, inspiration, and patience over many years.

We have recently learned that Morisuke Hasumi independently

extended the methods and results of our theses [26] to obtain the extended Cauchy-Read theorems of chapter 6 and the extended Beurling-Rubel-Shields theorem of chapter 7, while the final draft of this paper was in preparation, that is a few months after we first obtained these results. But in addition, he succeeded in characterizing the closed (weak* closed if $p = \infty$) subspaces of L^p on the Martin boundary of an admissible surface, which are invariant under multiplication by all bounded analytic functions! (C. f. M. Hasumi, Invariant Subspaces on Open Riemann Surfaces, Ann. Inst. Fourier Grenoble to appear).

Chapter 1

Preliminaries

1.1. Let $\mathbb{Z}$, $\mathbb{R}$, and $\mathbb{C}$ be, respectively, the integers, the reals, and the complex plane. Let $\Delta(a,r)$ be the open disk of radius r centered at a. Throughout this paper, R shall always be a hyperbolic Riemann surface, and $G(a,z)$ shall be the Green's function for R with pole at a. We shall let a_0 be a fixed point of R.

We shall introduce several classes of complex valued harmonic functions analogous to $H^p(R)$. For $1 < p < \infty$, $h^p(R)$ shall denote the class of functions f harmonic on R such that $|f|^p$ has a harmonic majorant, and $h^\infty(R)$ shall denote the class of bounded harmonic functions on R. We shall reserve the symbol $h^1(R)$ for later use. However, $hP(R)$ shall denote the class of functions f harmonic on R such that $|f|$ has a harmonic majorant.

Throughout this paper, L. H. M. shall stand for <u>least harmonic majorant</u>, and G. H. M. shall stand for <u>greatest harmonic minorant</u>. If we norm $h^p(R)$ by setting

$$\text{(1)} \qquad ||f|| = (\text{L. H. M. } |f|^p \Big|_{a_0})^{1/p} ,$$

then $h^p(R)$ becomes a Banach space (under pointwise algebraic operations) for $1 < p < \infty$. The spaces $h^\infty(R)$ and $hP(R)$ are also Banach spaces under, respectively, the supremum norm and the norm

$$\text{(2)} \qquad ||f|| = \text{L. H. M. } |f| \Big|_{a_0} .$$

We shall refer to these norms as the $h^p(R)$ norms, even if $p = 1$. The

Received by the editors September 17, 1973
Accepted after revisions November 7, 1974

proof that the $h^p(R)$ (respectively $hP(R)$) are Banach spaces is entirely analogous to the proof that the corresponding Hardy classes are Banach spaces (c. f. [17, chapter 1]). We note that if $R = \Delta(0, 1)$, then the $h^p(\Delta(0, 1))$ norm coincides with the usual norm on $H^p(\Delta(0, 1))$.

For $1 < p \leq q \leq \infty$, $h^q(R) \subseteq h^p(R)$ and $h^p(R) \subseteq hP(R)$. Consider, for example, the second inclusion. Let $f \in h^p(R)$. Clearly

$$|f| \leq (\text{L. H. M. } |f|^p)^{1/p} .$$

Due to the concavity of the function $x^{1/p}$, the right hand side is superharmonic. Thus $|f|$ admits a harmonic majorant, in fact

$$\text{L. H. M. } |f| \leq (\text{L. H. M. } |f|^p)^{1/p} .$$

Thus the $h^p(R)$ norm majorizes the $h^1(R)$ norm, and a similar argument shows that the $h^p(R)$ norms increase with increasing p.

The interior, boundary, and closure of a subset A of a topological space $(X, \mathfrak{J})$ shall be denoted, respectively, by

$$\text{Cl}_{\mathfrak{J}} A = \text{Cl}_X A = \text{Cl} A$$

$$\text{Int}_{\mathfrak{J}} A = \text{Int}_X A = \text{Int} A$$

$$\partial_{\mathfrak{J}} A = \partial_X A = \partial A$$

Suppose $(X, \mathfrak{J})$ is locally compact, and let $C_b(X)$ denote the bounded, complex valued, continuous functions on X. Let ∞ be the point adjoined in the Alexandroff compactification of X, and let $C_0(X)$ denote the complex valued functions f, continuous on X, such that $\lim_{x \to \infty} f(x) = 0$. We shall need the β, or strict topology on $C_b(X)$ introduced by Buck [4]. For each non-negative $k \in C_0(X)$, let ρ_k be the seminorm

$$\rho_k(f) = \sup \{k(x) |f(x)| : x \in X\} , \quad f \in C_b(X) . \tag{3}$$

The β topology is the unique locally convex topology on $C_b(X)$ generated by the seminorms ρ_k, $k \in C_0(X)$, $k \geq 0$.

For future reference, if f is complex valued on X, and if

(4) $$\sup \{k(x) \, |f(x)| : x \in X\} < \infty ,$$

for each non-negative $k \in C_0(X)$, then f is bounded on X. The proof may be carried out via a variant of a construction by Rubel and Shields [35].

1.2. Following Rubel and Shields [35], we shall call a function $f \in H^\infty(R)$ interior if $f \not\equiv 0$, and

(1) $$\mathrm{Cl}_\beta \, f\cdot H^\infty(R) = f\cdot H^\infty(R).$$

We shall say that $f \in H^\infty(R)$ is exterior if

(2) $$\mathrm{Cl}_\beta \, f\cdot H^\infty(R) = H^\infty(R).$$

The Rubel-Shields theorem describes the β closed ideals of $H^\infty(\Delta(0, 1))$. It asserts that each such ideal is principal, and in fact is of the form

$$f\cdot H^\infty(\Delta(0, 1)),$$

where f is an interior function determined uniquely, modulo multiplication by invertible elements in the algebra $H^\infty(\Delta(0, 1))$, by the ideal [35]. Suggestively, the statement of the Rubel-Shields theorem involves only quantities defined on the interior of the unit disk rather than on the boundary. Thus the statement may be carried over without change to arbitrary R. This work began as an attempt to generalize the Rubel-Shields theorem to $H^\infty(R)$. In keeping with the spirit of their paper, we have used interior methods whenever possible. Our methods are as heavily dependent on function theory and potential theory as on abstract harmonic analysis. We have also been heavily influenced toward an interior approach by the work of Maurice Heins, (c. f. [16], [17]). We hope this classical viewpoint will be refreshing.

As indicated in the introduction, the main theorem of this memoir is a generalization of the Beurling-Rubel-Shields theorem to $H^p(R)$, for R admissible. Although we have succeeded in completely characterizing the β closed ideals of $H^\infty(R)$, our characterization does not say that each such ideal is principal. In fact, we shall show by example that for some admissible R, there exist non-principal β closed ideals in $H^\infty(R)$. Similar comments apply to $H^p(R)$ for $1 \leq p < \infty$. A related result is that some $f \in H^\infty(R)$ may fail to have an interior-exterior factorization. Rubel and Shields have recently given examples of similar phenomena in several variables [36].

We shall provide examples, based on a Rudin type bubble region [39], showing that our main theorem cannot be carried over to $H^\infty(R)$ for arbitrary R, even if R admits lots of bounded analytic functions. We unfortunately have not obtained necessary and sufficient conditions on R for our generalization to hold.

As we mentioned in the introduction, we recently learned that M. Hasumi independently extended the methods and results of our thesis [26] to obtain the extended Beurling-Rubel-Shields theorem of chapter 7, a few months after we first obtained these results. In addition, he succeeded in characterizing the closed (weak* closed if $p = \infty$) subspaces L^p on the Martin boundary, which are invariant under multiplication by all bounded analytic functions! (C. f. M. Hasumi, <u>Invariant Subspaces on Open Riemann Surfaces</u>, Ann. Inst. Fourier Grenoble, to appear.) Hasumi's remarkable result immediately yields a characterization of the closed (β closed if $p = \infty$) supspaces of $h^p(R)$ invariant under multiplication (via the product $<,>$ introduced in chapter 3) by all functions in $H^\infty(R)$, for admissible surfaces R. Hasumi's definition of an admissible surface is formally weaker than ours. The question of the equivalence of the two definitions remains open.

Chapter 2

L. A. M.'s, Inner-Outer Factorizations

2.1. In this chapter, we shall consider the inner-outer factorization of meromorphic functions of bounded characteristic on a Riemann surface R from a potential-theoretic point of view. Essentially, we shall employ Parreau's decomposition of a positive harmonic function into singular and quasi-bounded parts. [29]. To carry out this program, we shall need to consider several vector lattices of harmonic functions, and several direct sum decompositions of these vector lattices.

Let $HP(R)$ denote the real subspace of real valued harmonic functions lying in $hP(R)$. The space $HP(R)$ provides a familiar example of a Banach lattice, when endowed with the pointwise order relation [5, chapter 2]. The lattice supremum is simply

$$f \vee g = \text{L. H. M.} \max(f, g), \quad f, g \in HP(R) \quad ,$$

where $\max(f, g)$ denotes the pointwise supremum.

Let $Q(R)$ be the band of quasi-bounded functions in $HP(R)$, and let $I(R)$ be the band of singular harmonic functions in $HP(R)$. These notions are due to Parreau, as is the direct sum decomposition

(1) $$HP(R) = I(R) \oplus Q(R) \quad ,$$

([29], c.f. also [5, chapter 2] or [17, pp. 6-9]). Let $HB(R)$ be the Banach space of real valued functions in $h^\infty(R)$. The band $Q(R)$ is simply the band in $HP(R)$ generated by $HB(R)$, and $I(R)$ is $HB(R)^\perp$, the band orthogonal to $HB(R)$ (and $Q(R)$) . Our vector lattice terminology will always be borrowed from [6] and [41, chapter V .

Call a function $f \in hP(R)$ <u>quasi-bounded</u> if $\text{Re } f$ and $\text{Im } f$ are both quasi-bounded. Clearly, $f \in hP(R)$ is quasi-bounded if and only if $\text{L. H. M. } |f| \in Q(R)$. We shall denote the normed linear subspace of quasi-bounded functions in $hP(R)$ by $h^1(R)$.

Theorem 1: $h^1(R)$ is a Banach space.

Proof: It suffices to prove $Q(R)$ is a complete subspace of $HP(R)$. This is an immediate consequence of the fact that $Q(R)$ is a band in the Banach lattice $HP(R)$ [6, p. 100], [41, p. 235].

2.2. We require an extension of $HP(R)$ which is no longer a Banach space, but which is a vector lattice.

Definition 1: $SP(R)$ is the set of all differences of positive functions, harmonic on R except for isolated singularities.

Thanks to positivity, each singularity of an $SP(R)$ function is either removable or logarithmic [16, p. 58]. Thus each function in $SP(R)$ may be extended by continuity to all of R, as an extended real valued function. We shall always suppose this extension has been made. Then $SP(R)$ becomes an ordered vector space, under vector space operations and ordering defined pointwise away from singularities, and by continuity at singularities. We shall take the liberty of also denoting the resulting ordered vector space by $SP(R)$.

Theorem 1: $SP(R)$ is an order complete vector lattice.

The reader should note that each function in $SP(R)$ is a difference of positive superharmonic functions. However, theorem 1 is not simply a corollary to the Riesz representation theorem [5], because $SP(R)$ is more restrictive than the lattice of differences of positive superharmonic functions.

Proof: For $f \in SP(R)$, let $c_f(z)$ denote the order of the singularity of f at z. Set $c_f(z) = 0$ if f has no singulairty at z. It suffices to consider only the infima of sets of positive functions in $SP(R)$ [41, theorem V.1.2]. Let V be such a set, and let $0 \leq h \leq f$ for each f in V. By the minimal property of the Green's function, for $f \in V$ and $w \in R$, $f(z) - c_f(w)G(w, z)$ is the largest positive

$SP(R)$ function no larger than f, and harmonic at w. Applying the same argument to h, we have

$$0 \leq h(z) - c_h(w)G(w, z) \leq f(z) - c_f(w)G(w, z).$$

Thus

$$0 \leq h(z) - \sum(c_h(w)G(w, z): \ c_h(w) \neq 0)$$

$$\leq f(z) - \sum(c_f(w)G(w, z): \ c_f(w) \neq 0),$$

and by the Harnack monotone convergence theorem, both functions are harmonic on R. Since $0 \leq c_h(w) \leq c_f(w)$, we have

$$0 \leq \sum(c_h(w)G(w, z): \ c_h(w) \neq 0)$$

$$\leq \sum(c_f(w)G(w, z): \ c_f(w) \neq 0),$$

and both functions are in $SP(R)$. Let

(1) $$g_1(z) = \bigwedge(f(z) - \sum(c_f(w)G(w, z): \ c_f(w) \neq 0); \ f \in V\ ))$$

$$g_2(z) = \sum(c(w)G(w, z): \ c(w) \neq 0),$$

where $c(w) = \inf(c_f(w): \ f \in V)$. The infimum $\bigwedge$ is taken in $HP(R)$, and the second infimum is taken in R. Clearly, g_1 and $g_2 \in SP(R)$, and $h \leq g_1 + g_2 \leq f$ for each $f \in V$. Thus $g_1 + g_2$ is the lattice infimum of V.

Let $G(R)$ denote the vector space of potentials lying in $SP(R)$. Clearly, $G(R) \subseteq HP(R)^\perp$. From the general theory of vector lattices,

$$SP(R) = HP(R)^\perp \oplus HP(R)^{\perp\perp} .$$

Since $G(R)$ and $HP(R)$ generate $SP(R)$ as a vector space, $G(R) = HP(R)^\perp$, $HP(R) = G(R)^\perp$, and both $G(R)$ and $HP(R)$ are bands in $SP(R)$. In particular, the lattice supremum of a subset of $HP(R)$ is the same, whether taken in the lattice $HP(R)$ or $SP(R)$. Throughout this paper, $\bigvee$ and $\bigwedge$ shall indicate, respectively, the lattice supremum and the lattice infimum in $SP(R)$.

In addition we have proved

Theorem 2: $\mathcal{G}(R)$, $I(R)$, and $Q(R)$ are mutually orthogonal bands in $SP(R)$, and

(2)
$$SP(R) = \mathcal{G}(R) \oplus I(R) \oplus Q(R),$$

where the associated projection maps are positive.

2.3. It is instructive to consider the direct sum decomposition 2.2.2 in case $R = \Delta(0, 1)$. The decomposition states that $h \in SP(\Delta(0, 1))$ may be uniquely written as the sum

(1)
$$h(z) = \sum c_i \ln \left| \frac{1-\bar{a}_i z}{a_i - z} \right| + \int_0^{2\pi} K(z, e^{i\theta})\ d\mu_s(e^{i\theta}) + \int_0^{2\pi} K(z, e^{i\theta})\ f(e^{i\theta}) \frac{d\theta}{2\pi}, \quad z \in \Delta(0, 1),$$

where $f \in L^1(d\theta)$ and is the Fatou radial limit function of h, $d\mu_s$ is a regular Baire measure singular with respect to Lebesgue measure, $K(z, e^{i\theta})$ is the Poisson kernel for $\Delta(0, 1)$, a_i, $i = 1, 2, \cdots$, is a univalent enumeration of the singularities of h, and c_i is the order of the singularity of h at a_i. It is standard that if h is harmonic, the last two terms of the right hand side of equation 1 correspond to the singular-quasi-bounded decomposition of h. Nevertheless, it may be useful to briefly indicate why this is so. Suppose $h \geq 0$ and let f be the Fatou radial limit function of h. Consider the Poisson integrals

$$h_n(z) = \int_0^{2\pi} K(z, e^{i\theta}) \min (f(e^{i\theta}), n) \frac{d\theta}{2\pi}, \quad z \in \Delta(0, 1),$$

for $n = 1, 2, \ldots$ Clearly,

$$\lim_{n \to \infty} h_n = \bigvee \{g \in HB(R): g \leq h\},$$

so $\lim_{n \to \infty} h_n$ is the quasi-bounded part of h. By the Lebesgue monotone convergence theorem, $\lim_{n \to \infty} h_n$ is the Poisson integral of f. To see why the middle term of 1 is the singular part of h, recall that h may be written as the Poisson integral of a unique regular Baire measure $d\mu$, and that $f(e^{i\theta})d\theta$ is the absolutely continuous part of $d\mu$ [22].

Suppose that g is an analytic function of bounded characteristic on $\Delta(0, 1)$, and let $|g| = \exp h$. Then $h \in SP(\Delta(0, 1))$ and the decomposition 1 corresponds to the factorization of f into a product of a Blaschke product, a singular function, and an outer function. This observation motivates the following definition for hyperbolic Riemann surfaces R.

Definition 1: A locally meromorphic modulus, or l. m. m., is an extended real valued function F, defined on R, such that for each $p \in R$, there exists an open neighborhood V of p such that $F = |f|$ for some function f meromorphic on V. A locally analytic modulus, or l. a. m. is defined similarly. The l. m. m. F is of bounded characteristic if $\ln F \in SP(R)$, a generalized Blaschke product if $\ln F \in G(R)$, singular if $\ln F \in I(R)$, inner if $\ln F \in G(R) \oplus I(R)$, and outer if $\ln F \in Q(R)$. The corresponding notions for functions f meromorphic on R are defined in terms of the l. m. m. $|f|$.

From theorem 2.2.2., we immediately conclude

Theorem 1: Each l. m. m. of bounded characteristic on R has unique inner-outer factorization, and a unique generalized Blaschke product-singular-outer factorization.

We note that an l. a. m. is a special kind of PL function (c. f. [17, p. 29]). Clearly the set of l. m. m.'s on R, as well as each of the classes of l. m. m.'s occuring in definition 1, are preserved under the formation of products and quotients. Of course, the following exceptions must be noted: In no case can the l. m. m. in the denominator vanish identically, and the class of l. a. m.'s is not preserved under the formation of quotients.

It is trivial to show that the meromorphic functions of bounded characteristic form a field, and that they coincide with the Lindelöfian meromorphic functions introduced by Heins [15, pp. 440-443]. The l. m. m.'s of bounded characteristic are the set of quotients of bounded l. m. m.'s. The work of Heins shows that for some R, there are analytic functions of bounded characteristic which are not quotients of bounded analytic functions [15, p. 444]. The work of Sarason and Voichick shows that there are bounded analytic functions on annuli with no inner-outer factorization, if the factors are required to be analytic functions [40], [45]. Finally, $H^p(R)$ is a subset of the set of meromorphic functions of bounded characteristic, as may be seen from the inequality $\ln|f| \leq$ L. H. M. $|f|$.

2.4. There is a close connection between l. m. m.'s and multiplicative multiple valued meromorphic functions. The connection is best approached via an analytic continuation process. To simplify the exposition, all surfaces and bordered surfaces will be connected throughout this paper. In addition, we shall define <u>simply connected</u> to also mean connected.

Let U be an open subset of R, and let F be an l. m. m. on R. We shall say that a meromorphic function f with domain U is a <u>branch</u> of F on U if $F = |f|$ on U. If $U = R$, f will be called a <u>global branch</u> of F.

Let (p, f) be a <u>meromorphic function element</u> on R, that is, let f be a meromorphic function with domain a simply connected open neighborhood U of p, where $p \in R$. We shall say (p, f) <u>belongs</u> to the l. m. m. F if f is a branch of F on U. Clearly, if (p, f) belongs to F, we may meromorphically continue (p, f) along any arc in R by choosing suitable function elements from F. That is, given an arc $\gamma(t)$, $0 \leq t \leq 1$, in R, we may find a chain of function elements in F,

(1) (p_i, f_i), $i = 0, 1, 2, \ldots, n$,

satisfying the following: For $i = 0, 1, 2, \ldots, n$, $p_i = \gamma(t_i)$, where the sequence t_i, $i = 0, 1. \ldots, n$, in non-decreasing, $t_0 = 0$, and $t_n = 1$. Further, letting U_i be the domain of f_i for each i, we have $\{\gamma(t)\colon t_i \leq t \leq t_{i+1}\} \subseteq U_i$ for $i = 0, \ldots, n-1$, and $\gamma(t_n) \in U_n$. Finally, we require that $f_i = f_{i+1}$ on the component of $U_i \cap U_{i+1}$ containing p_i for $i = 0, 1, \ldots, n-1$.

There are three important consequences to this. First, if V is any simply connected region in R, the monodromy theorem shows that F has a branch on V [1, pp. 32-34]. Second, from this observation, it is easy to show that if (q, g) is any meromorphic continuation of a function element (p, f) belonging to F, then (q, g) also belongs to F. Thus we may think of F as the modulus of a multiplicative multiple valued meromorphic function in the sense of Weierstrass, that is, as the modulus of the meromorphic analogue of an analytic continuation class [48], [49]. Third, let $(\Delta(0, 1), \psi)$ be the universal covering surface of R, and let T be the group of covering transformations. Then there exists a meromorphic function f on $\Delta(0, 1)$ such that $|f|$ is automorphic and $|f| = F \circ \psi$.

Conversely, given a meromorphic function f on $\Delta(0, 1)$ such that $|f|$ is automorphic, there exists an l. m. m. F on R such that $|f| = F \circ \psi$. Voichick called functions such as f modulus invariant, and he and Sarason independently introduced multiple valued inner-outer factorizations via the modulus invariant functions, and the inner-outer factorization of meromorphic functions of bounded characteristic on the unit disk [40], [45]. To reconcile the Sarason-Voichick approach with ours, we must investigate the behavior of $SP(R)$ when lifted to the universal covering surface of R.

Suppose h is an extended real valued function on R such that $h \circ \psi \in SP(\Delta(0, 1))$. Let $s = h \circ \psi \vee 0$. For $\tau \in T$, $s \circ \tau$ and $s \circ \tau^{-1}$ majorize $h \circ \psi$ and 0, so $s \circ \tau$ and $s \circ \tau^{-1}$ majorize s. Thus s is automorphic, and consequently $h \in SP(R)$. A simple extension of this argument shows that the map

$$\alpha : h \longrightarrow h \circ \psi$$

carrying SP(R) into SP(Δ(0, 1)) preserves suprema, and hence all lattice operations. More generally, a similar argument establishes the following lemma:

Lemma 1: Let $A \subseteq SP\Delta(0, 1)$ be such that $A \circ \tau \subseteq A$ for each $\tau \in T$. Then if $\bigvee A$ exists in SP(Δ(0, 1)), it is automorphic.

Now let h be a positive function in SP(R) and let $h \circ \psi = s+t$ be the direct sum decomposition of $h \circ \psi$ into a generalized Blaschke product and an HP(Δ(0, 1)) function. Then

$$s = \bigvee\{u \in G(\Delta(0, 1)) : u \leq h \circ \psi\}$$

$$t = \bigvee\{u \in HP(\Delta(0, 1)) : u \leq h \circ \psi\} .$$

By lemma 1, both s and t are automorphic, and so $s = \alpha(g)$, $t = \alpha(f)$, for some f and g in SP(R). But $\alpha(g \wedge u) = \alpha(g) \wedge \alpha(u) = 0$ for each positive $u \in HP(R)$, since $G(\Delta(0, 1)) = HP(\Delta(0, 1))^{\perp}$. Since α is 1-1, $g \in HP(R)^{\perp} = G(R)$. Similarly $f \in HP(R)$. By appealing to the direct sum decomposition 2.1.1 of f and applying a similar argument, we obtain

Theorem 1: Let $h = g + s + u$ be the direct sum decomposition 2.2.2 of $h \in SP(R)$. Then $h \circ \psi = g \circ \psi + s \circ \psi + u \circ \psi$ is the direct sum decomposition 2.2.2 in SP(Δ(0, 1)) of $h \circ \psi$.

As a corollary, we have the well known result that $z \rightarrow G(a, \psi(z))$ is a sum a Green's functions of Δ(0, 1) [25, pp. 334-338]. We also have

Theorem 2: Let $f = BSF$ be the generalized Blaschke product-singular-outer factorization of an l. m. m. f of bounded characteristic on R. Let $\hat{f}$, $\hat{B}$, $\hat{S}$, and $\hat{F}$ be meromorphic functions on Δ(0, 1) such that $|\hat{f}| = f \circ \psi(z)$, $|\hat{B}| = B \circ \psi(z)$, etc. Then $\hat{f} = \lambda\hat{B}\hat{S}\hat{F}$ is the Blaschke product-singular-outer factorization of $\hat{f}$ in the field of meromorphic functions of bounded characteristic on Δ(0, 1).

Consequently, the factorization guaranteed by theorem 2.3.1 is essentially the same as the Sarason-Voichick factorization.

2.5. By passing to the universal covering surface and using standard results for the disk, we could easily show that each $h \in h^p(R)$ is quasi-bounded if $p > 1$, as is each $f \in H^1(R)$. However, we prefer an internal approach. Heins has summarized a series of incredibly clever arguments involving subharmonicity which have these results, together with the Szëgo maximal theorem, as consequences [17, chapter 2]. We shall say that a function α satisfies the de la Valée Poussin condition if α is a continuous, non-decreasing, non-negative, extended real valued function on $[-\infty, +\infty]$, which is convex on $\mathbb{R}$, and which satisfies

$$\lim_{x \to +\infty} \alpha(x)/x = +\infty .$$

Let α be such a function. Heins showed that if u is subharmonic on R and $\alpha \circ u$ admits a harmonic majorant, then the following is true:

Theorem 1: The functions $\max(u, 0)$ and $\alpha(\max(u, 0))$ admit quasi-bounded harmonic majorants. If $u \not\equiv -\infty$, then u admits a unique representation of the form

$$u = h - s - g ,$$

where $h \in Q(R)$, $s \in I(R)$ and $s \geq 0$, and g is a non-negative superharmonic function such that G. H. M. $g = 0$. Furthermore, $\alpha \circ h$ has a harmonic majorant, and

$$\text{L. H. M.}\, \alpha \circ u = \text{L. H. M.}\, \alpha \circ (\text{L. H. M.}\, u) = \text{L. H. M.}\, \alpha \circ h .$$

Heins called this result the extended theorem of Szegö - Solomentsev. We shall have a variety of uses for it.

Let $1 < p < \infty$ and let $h \in h^p(R)$. The function $|h|$ is subharmonic. Set

$$\alpha(x) = (\max(x, 0))^p$$

and apply the extended theorem of Szegö-Solomentsev to show that L. H. M. $|h| \in Q(R)$. We have just proved

<u>Theorem</u> 2: Each function in $h^p(R)$ is quasi-bounded for $1 \leq p \leq \infty$. Equivalently, $h^p(R) \subseteq h^1(R)$ for $1 \leq p \leq \infty$.

Let f be an l. a. m. admitting a harmonic majorant. Then ln f is subharmonic. Write

$$\ln f = h - s - g \quad , \tag{1}$$

where $h \in Q(R)$, $s \in I(R)$, and $g \in G(R)$. We can use the uniqueness of the Riesz decomposition of a difference of positive superharmonic functions together with theorem 1 to conclude immediately that s and g are non-negative. Or, we can argue as follows: Let $\alpha(x) = \exp x$. By the extended Szegö-Solomentsev theorem, $\exp(\ln^+(f))$ has a quasi-bounded harmonic majorant, say H. Then $\ln f - H \leq 0$, so $h - H \leq 0$, $s \geq 0$, and $g \geq 0$ by the positivity of the projection maps associated with the direct sum decomposition 2.2.2. In particular, the inner factor of f is bounded above by 1. Also, since $0 \leq$ L. H. M. $f \leq H$, L. H. M. $f \in Q(R)$. Thus we have shown the first part of

<u>Theorem</u> 3: Let f be an l. a. m. on R admitting a harmonic majorant. Then L. H. M. f is quasi-bounded, and the inner factor of f is bounded above by 1. If f is actually the modulus of an analytic function u defined on R, then u is quasi-bounded.

Let f be an l. a. m. such that f^p has a harmonic majorant for some $p \in [1, \infty)$, and let F be the outer factor of f. Then F^p admits a quasi-bounded harmonic majorant, and

$$\text{L. H. M. } f^p = \text{L. H. M. } F^p \quad .$$

If f is a bounded l. a. m., then

$$\sup\{f(z) : z \in R\} = \sup\{F(z) : z \in R\} \quad .$$

<u>Proof</u>: Suppose f^p admits a harmonic majorant, where $1 \leq p < \infty$. Let

$\alpha(x) = \exp px$ and apply the Szegö-Solomentsev theorem to $\ln f$. Thanks to the non-negativity of s and g in equation 1,

$$L.\ H.\ M.\ f^p = L.\ H.\ M.\ F^p \quad ,$$

since $F = \exp h$.

Suppose that f is bounded. Then

$$\sup\{\ln f(z) : z \in R\} = \ln m,$$

where m is the supremum norm of f on R. Thus $k = \sup\{h(z) : z \in R\} \geq \ln m$, since s and g are non-negative. However, $\ln f - \ln m \leq 0$, so $h - \ln m \leq 0$ since $h - \ln m$ is the quasi-bounded component of $\ln f - \ln m$. Thus $k = \ln m$, and consequently f and F have the same supremum norm on R.

The last two assertions of theorem 3 correspond to the well known result that inner functions in $H^\infty(\Delta(0, 1))$ act as $H^p\ (\Delta(0, 1))$ isometries via multiplication. The relation of these assertions to Heins' results on P. L. functions and the Szegö maximal theorem should be self evident (c. f. [17, theorems 2.5 and 2.6]). We shall use theorem 3 to establish a generalization of the Szegö maximal theorem in section 4.3.

Chapter 3

The Banach Algebra $h^\infty(R)$

3.1. Our main goal in this chapter and the next is to develop an abstract boundary for R, without using anything more than the Gelfand-Neumark theorem, Stokes theorem, and a few simple facts about harmonic majorants. In this chapter, we shall use the inner product structure of $h^2(R)$ to obtain a multiplication, which we shall denote by $<\ ,\ >$, on $h^\infty(R)$. We shall show that $h^\infty(R)$ is a B^* algebra under the multiplication $<\ ,\ >$, and we shall show that the maximal ideal space M of $h^\infty(R)$ is actually the Hayashi boundary of R.

Because we shall use the Stokes-Green theorem, a few simple comments about conjugate differentials and Jordan subregions will be helpful. If ω is a differential 1-form on R, we shall denote the conjugate differential by $*\omega$. The conjugate differential operator to d shall be denoted by $*d$ (c. f. [42], [43, chapter 6]). Fortunately, we shall not explicitly need the wedge product of the exterior differential calculus, so no confusion can arise with the lattice infimum in SP(R).

We shall assume the standard elementary facts and terminology pertaining to bordered Riemann surfaces, as contained, say, in [1, chapter II]. If $S \cup \Gamma$ is a bordered Riemann surface, we shall always assume the border Γ is oriented positively with respect to the interior S. If ω is a continuous differential 1-form on Γ, we shall sometimes speak of ω as a measure. By that, we shall of course mean the regular Baire measure induced, via the Riesz-Markov theorem, by the continuous linear functional

$$f \longrightarrow \int_\Gamma f\omega , \quad f \in C(\Gamma) .$$

Here, $C(\Gamma)$ is the space of continuous complex valued functions on Γ with compact support

Suppose that $S \cup \Gamma$ is compact and that h is harmonic on $S \cup \Gamma$, except possibly for isolated logarithmic singularities, none of which occur on Γ. Suppose that u is harmonic on $S \cup \Gamma$. A familiar consequence of the Stokes-Green theorem is

$$(1) \qquad \int_{\Gamma} (u_{*}dh - h{*}du) = -2\pi \sum c(\xi)u(\xi),$$

where the sum on the right is taken over the singularities of h, and $c(\xi)$ is the order of the singularity of h at ξ. A simple and familiar consequence of 1 is that

$$\frac{1}{-2\pi}{*}dG_S(a,z) = \frac{1}{-2\pi}{*}d(z \to G_S(a,z))$$

is harmonic measure on Γ with respect to a, where $G_S(a, z)$ is the Green's function for S. The reader is referred to [5, chapter 3] for a complete discussion of harmonic measure.

Throughout this paper, a <u>Jordan subregion of R</u> shall be more than a region of R bounded by disjoint simple closed curves. A region S of R shall be a Jordan subregion of R if $\mathrm{Cl}S$ may be endowed with a conformal structure making $\mathrm{Cl}S$ into a compact bordered Riemann surface with border ∂S. In addition, we require that this conformal structure be consistent with the topology on $\mathrm{Cl}S$, and the conformal structure on S, induced by R. The closure of a Jordan subregion shall be called a <u>closed Jordan subregion</u>.

Clearly, this definition would make sense even if R were parabolic, so we shall also speak of Jordan subregions of parabolic surfaces. If S is a bounded region of $\mathbb{C}$, bounded by finitely many disjoint Jordan curves, it is easy to show that S is a Jordan subregion of $\mathbb{C}$ [26, section 2.3]. The proof involves a simple application of the Caratheodory-Osgood-Taylor theorem [16, chapter 6].

It is standard that R may be exhausted by a sequence of Jordan subregions [1, section II.12]. We shall let J denote the directed set of Jordan subregions of R, containing a_0, ordered by inclusion. For each $S \in J$, we shall denote the Green's function of S by $G_S(a, z)$.

3.2. We are now ready to impose a Banach algebra structure on $h^\infty(R)$. The construction depends on a few simple facts about harmonic majorants. From the equation

$$\text{(1)} \qquad \text{L. H. M.}\, |u|^2 \Big|_a = \lim_{S \in J} \frac{1}{-2\pi} \int_{\partial S} |u(z)|^2 {*dG_S(a,z)} \ ,$$

for all $u \in h^2(R)$, we obtain the existence of the limit

$$\text{(2)} \qquad \lim_{S \in J} \frac{1}{-2\pi} \int_{\partial S} u(z)v(z){*dG_S(a,z)} \ ,$$

for all u and $v \in h^2(R)$ and each $a \in R$. It suffices to show that 2 exists for u and v real valued. In that case, 2 is simply

$$\text{(3)} \qquad \tfrac{1}{4}(\text{L. H. M.}|u+v|^2 - \text{L. H. M.}|u-v|^2)\Big|_a$$

Thus we may make the following definition:

<u>Definition</u> 1:

$$< f,g > \Big|_a = \lim_{S \in J} \frac{1}{-2\pi} \int_{\partial S} fg{*dG_S(a,z)} \ ,$$

for all $(f,g) \in h^\infty(R) \times h^\infty(R)$ and each $a \in R$.

Clearly $< \ , \ >$ is bilinear, and $< f,g > \geq 0$ if f and g are non-negative. Consequently, the function $f \to < f,g >$ is increasing on $HB(R)$ if $g \geq 0$. Furthermore, we have

Lemma 1: The bilinear map $< \ , \ >$ maps $h^\infty(R) \times h^\infty(R)$ into $h^\infty(R)$.

Proof: Let f and $g \in h^\infty(R)$. The harmonicity of $< f,g >$ follows from 3. The boundedness of $< f,g >$ follows from

$$(4) \qquad \left| \frac{1}{-2\pi} \int_{\partial S} fg{*}dG_S(a,z) \right| \le ||f||_\infty ||g||_\infty \left(\frac{1}{-2\pi} \int_{\partial S} {*}dG_S(a,z)\right) ,$$

for all $S \in J$ containing a. The integral on the right hand side is simply the harmonic function $\equiv 1$ on S, evaluated at a, and so equals 1. Here, $|| \ ||_\infty$ denotes the $h^\infty(R)$ norm. We have almost completed the proof of

Theorem 1: $h^\infty(R)$ is a commutative B^* algebra with unit, when endowed with the $h^\infty(R)$ norm, with complex conjugation as an involution, and with $< \ , \ >$ as a multiplication.

Proof: From 4, $||< f,g >||_\infty \le ||f||_\infty ||g||_\infty$ for all $f,g \in h^\infty(R)$. The product $< \ , \ >$ is clearly commutative, and thanks to the bilinearity of $< \ , \ >$, $h^\infty(R)$ is a (possibly non-associative) algebra. The function $\equiv 1$ on R is clearly a unit element, and, of course, we already know $h^\infty(R)$ is a Banach space under the $h^\infty(R)$ norm. It remains to show that $< \ , \ >$ is associative. For this, we need the following lemma:

Lemma 2: For each $S \in J$, let $|| \ ||_S$ denote the $h^2(S)$ norm, and let $|| \ ||$ denote the $h^2(R)$ norm. Then for each $S \in J$,

$$\frac{1}{-2\pi} \int_{\partial S} |< f,g > - fg| {*}dG_S(a_0,z) \le (||f||^2 - ||f||_S^2)^{1/2} (||g||^2 - ||g||_S^2)^{1/2} ,$$

for all $f,g \in h^\infty(R)$.

Proof: Consider the equations

$$\text{L. H. M. } |f-\bar{g}|^2 = < f-\bar{g}, \bar{f}-g > = \text{L. H. M. } |f|^2 + \text{L. H. M. } |g|^2 - 2\text{Re} < f,g >$$

and

$$|f-\bar{g}|^2 = |f|^2 + |g|^2 - 2\mathrm{Re} fg .$$

Since L. H. M. $|f-\bar{g}|^2 - |f-\bar{g}|^2 \geq 0$, we have

$$(5) \qquad \mathrm{Re}(< f,g > - fg)$$

$$\leq \frac{1}{2}(\text{L. H. M. } |f|^2 - |f|^2 + \text{L. H. M. } |g|^2 - |g|^2) ,$$

Inequality 5 still holds if f is replaced by λf in the left hand side, where λ is any constant of modulus 1. Therefore,

$$|< f,g > - fg| \leq \frac{1}{2}(\text{L. H. M. } |f|^2 - |f|^2 + \text{L. H. M. } |g|^2 - |g|^2)$$

Consequently, for each $S \in J$,

$$\frac{1}{-2\pi}\int_{\partial S} |< f,g > - fg| *dG_S(a_0,z) \leq \frac{1}{2}A + \frac{1}{2}B ,$$

where

$$A = \frac{1}{-2\pi}\int_{\partial S}(\text{L. H. M. } |f|^2 - |f|^2)*dG_S(a_0,z) = ||f||^2 - ||f||_S^2 ,$$

$$B = \frac{1}{-2\pi}\int_{\partial S}(\text{L. H. M. } |g|^2 - |g|^2)*dG_S(a_0,z) = ||g||^2 - ||g||_S^2 .$$

By homogeneity,

$$(6) \qquad \frac{1}{-2\pi}\int_{\partial S} |< f,g > - fg| *dG_S(a_0,z) < \frac{1}{2}\lambda^2 A + \frac{1}{2\lambda^2}B ,$$

for each non-zero $\lambda \in \mathbb{R}$. Taking the minimum of the right hand side of 6, we obtain the desired inequality.

The proof of lemma 2 is reminiscent of the proof of the CBS inequality. One might suspect that a result similar to lemma 2, but involving $h^p(R)$ norms rather than $h^2(R)$ norms, could be established by modeling the proof after the proof of the Hölder inequality. This does not seem to be possible because $\frac{1}{p}f^p + \frac{1}{q}g^q - fg$ is not necessarily superharmonic, even for non-negative f and g in $h^\infty(R)$, unless $p = 2$. Here $\frac{1}{p} + \frac{1}{q} = 1$.

For each $f \in h^\infty(R)$, indeed for each $f \in h^2(R)$,

$$(7) \qquad \|f\|_S \nearrow \|f\|, \quad S \in J.$$

Thus, by changing the base-point a_0 if necessary, we may conclude from lemma 2,

$$(8) \qquad \lim_{S\in J} \int_{\partial S} |< f,g > - fg| *dG_S(a,z) = 0,$$

for all $f,g \in h^\infty(R)$ and each $a \in R$. We now may complete the proof of theorem 1.

Proof: Let f,g, and $h \in h^\infty(R)$. Applying 8 twice, we obtain

$$(9) \qquad \lim_{S\in J} \frac{1}{-2\pi} \int_{\partial S} |< f,< g,h >> - fgh| *dG_S(a,z)$$

$$\leq \lim_{S\in J} \frac{1}{-2\pi} \int_{\partial S} |< f,< g,h >> - f< g,h >| *dG_S(a,z)$$

$$+ \|f\|_\infty \lim_{S\in J} \frac{1}{-2\pi} \int_{\partial S} |< g,h > - gh| *dG_S(a,z)$$

$$= 0.$$

A similar argument applied to $<< f,g >,h >$ shows that the value of both $< f,< g,h >>$ and $<< f,g >,h >$ at a is

$$(10) \qquad \lim_{S \in J} \frac{1}{-2\pi} \int_{\partial S} fgh{*}dG_S(a,z) ,$$

for each $a \in R$. To obtain 10, we must observe that 9 implies

$$\lim_{S \in J} \frac{1}{-2\pi} \int_{\partial S} < f,< g,h >>{*}dG_S(a,z)$$

$$= \lim_{S \in J} \frac{1}{-2\pi} \int_{\partial S} fgh{*}dG_S(a,z) ,$$

and that the left hand limit is simply $< f,< g,h >>$ evaluated at a, due to the harmonicity of $< f,< g,h >>$. This disposes of the associativity of $< \;,\; >$.

To show that $h^\infty(R)$ is a B^* algebra, we must verify that $||< h, \bar{h} >||_\infty = ||h||_\infty ||\bar{h}||_\infty$, for each $h \in h^\infty(R)$. But by 1,

$$(11) \qquad < h,\bar{h} > = \lim_{S \in J} \text{L. H. M. } |h|_S|^2 = \text{L. H. M. } |h|^2 .$$

For each $S \in J$, set $u_S = \text{L. H. M. } |h|_S|^2$.

By the maximum principle,

$$\sup\{u_S(z) : z \in S\} = (\sup\{|h(z)| : z \in S\})^2$$

since both sides equal $\sup\{|h(z)|^2 : z \in \partial S\}$. Thanks to the monotone convergence of the non-decreasing net u_S, $S \in J$, to $< h,\bar{h} >$, we have

$$||< h,\bar{h} >||_\infty = \sup_{S \in J} (\sup_{z \in S} |h(z)|)^2 = ||h||_\infty^2 .$$

Since $||\bar{h}||_\infty = ||h||_\infty$, we obtain the desired equality. Clearly, $\overline{< f,g >} = < \bar{f},\bar{g} >$ for all $f, g \in h^\infty(R)$, so $h^\infty(R)$ is a B^* algebra.

3.3 The product $<\ ,\ >$ has a simple interpretation in terms of the Martin boundary. Following the notation of [5, chapters 12-14], let Δ be the Martin boundary of R and consider the distinguished subset Δ_1. For $u \in hP(R)$, let $\hat{u}$ denote the modification of the Naim fine limit boundary function of u introduced in [5, chapter 14]. Then $\hat{u}$ is defined a.e. on Δ_1 with respect to harmonic measure, and the quasi-bounded component of u is the function

(1) $$a \longrightarrow \int_{\Delta_1} \hat{u}\, d\omega_a ,$$

where $d\omega_a$ is harmonic measure at a. Because $\hat{u}$ is essentially defined via cluster sets, it is not hard to show that

(2) $$\text{L. H. M. } |u|^2 \Big|_a = \int_{\Delta_1} |\hat{u}|^2 d\omega_a$$

for each $u \in h^2(R)$. From expression 3.2.3, we see that

(3) $$< f,g > \Big|_a = \int_{\Delta_1} \hat{f}\hat{g}\, d\omega_a$$

for all $f, g \in h^\infty(R)$. In fact, the B^* algebra $h^\infty(R)$ is isomorphic to $L^\infty(d\omega_{a_0})$, the algebra of essentially bounded $d\omega_{a_0}$ measurable functions defined a.e., under the essential supremum norm.

If $R = \Delta(0,1)$, we may replace Δ_1 by $\partial\Delta(0,1)$, $\hat{u}$ by u^*, the Fatou radial limit function of u, and $d\omega_a$ by $\frac{1}{2\pi}K(a,e^{i\theta})d\theta$, where $K(a,e^{i\theta})$ is the Poisson kernel for $\Delta(0,1)$. The preceeding remarks may be duplicated word for word, and we obtain,

(4) $$< f,g > \Big|_a = \int_0^{2\pi} f^* g^* K(a,e^{i\theta}) \frac{d\theta}{2\pi} .$$

Suppose now that R is once again an arbitrary hyperbolic Riemann surface, and let $(\Delta(0,1),\psi)$ be the universal covering surface of R. Let T be the group of covering transformations. As in the argument leading up to Lemma 2.4.1, we may easily show that

$$(5) \qquad (\text{L. H. M. } |f+g|^2 - \text{L. H. M. } |f-g|^2)\circ\psi$$
$$= \text{L. H. M. } |f\circ\psi+g\circ\psi|^2 - \text{L. H. M.} |f\circ\psi-g\circ\psi|^2 ,$$

for all $f, g \in h^2(R)$. From expression 3.2.3, we immediately conclude that

$$(6) \qquad <f,g>\circ\psi = <f\circ\psi, g\circ\psi> ,$$

for all $f, g \in h^\infty(R)$. This, together with 4 gives us another reasonably simple interpretation of $< , >$. In addition, it is easy to see that the B^* algebra $h^\infty(R)$ is isomorphic to the B^* algebra $L^\infty_T(d\theta/2\pi)$, the subalgebra of $L^\infty(d\theta/2\pi)$ consisting of the T automorphic functions.

It is worth explaining why we have chosen to develop the product $< , >$ ab initio, rather than starting from equations 3 or 6. There are three reasons: First, we wish to keep the exposition as self contained as economically possible, in order to maximize the accessibility of this paper. In particular, we do not want to presume a knowledge of the Martin boundary. Second, as we explained in the introduction, we aesthetically prefer interior methods. This preference certainly rules out 6, and possibly 3, as a starting point. Finally, even if we started from equation 3 or 6, we would still have to do nearly as much work in order to prove the Cauchy-Read theorems of chapter 6. By contrast, our approach, though slightly longer, has the great advantage of giving us several additional results of interest. These include a new construction of the Hayashi boundary, which will turn out to be simply the maximal ideal space of $h^\infty(R)$, and an interesting operational calculus for $h^\infty(R)$.

3.4. Throughout the remainder of this paper, M shall denote the maximal ideal space of $h^\infty(R)$. In this section, we shall show that M is identical with the Hayashi boundary, and we shall introduce harmonic measure on M. Let

$$\phi: h^\infty(R) \to C(M) \tag{1}$$

denote the Gelfand transform. Of course, C(M) is the B^* algebra of continuous, complex valued function on M. Both the real subalgebra of real valued functions in $h^\infty(R)$, HB(R), and the real subalgebra of real valued functions on M, $C(M, \mathbb{R})$, are Banach lattices when endowed with the pointwise order relation. Because involution in both cases is simply complex conjugation, ϕ carries HB(R) onto $C(M, \mathbb{R})$.

Theorem 1: In addition to preserving the algebraic, metric, and involutive structures of $h^\infty(R)$ and C(M), ϕ also preserves order on HB(R) and $C(M, \mathbb{R})$.

Due to the Gelfand-Neumark theorem, only the assertion about order needs proof. Before giving the proof, let us examine the consequences of this assertion. Because

$$\phi: HB(R) \longrightarrow C(M, \mathbb{R}) \tag{2}$$

is onto, ϕ is a lattice isomorphism, that is ϕ preserves infima and suprema. Since ϕ is a lattice isomorphism and an isometry, the set of extreme points of the positive face of the unit ball of $HB(R)'$ is just M. Consequently, the map 2 coincides with the Kakutani representation of HB(R) considered as an AM space with unit (c.f. [6, p. 103] or [41, p. 247]).

In particular, this shows that M coincides with the abstract boundary introduced by Hayashi [13]. In the terminology of Constantinescu and Cornea, M is the harmonic boundary of the Hayashi compactification [5, p. 98]. We shall not obtain any of Hayashi's results on fine limits of positive superharmonic functions with respect to M, nor shall we obtain any of his results linking M to classification problems for Riemann surfaces.

For each $a \in R$, the linear functional

$$f \longrightarrow \phi^{-1}(f)\Big|_a , \tag{3}$$

is continuous on $C(M, \mathbb{R})$ and thus, by the Riesz-Markov theorem, induces a unique regular Baire measure ω_a defined by

$$\int_M f d\omega_a = \phi^{-1}(f)\Big|_a , \quad f \in C(M) . \tag{4}$$

In fact, ω_a is positive since the linear functional 3 is, and

$$\int_M d\omega_a = 1(a) = 1 . \tag{5}$$

<u>Definition</u> 1: For each $a \in R$, the measure ω_a shall be called <u>harmonic measure</u> at a. The measure ω_{a_0} shall be denoted simply by ω.

Let $a \in R$. By the Harnack inequalities, there exist strictly positive finite constants λ and μ such that

$$\lambda \int f d\omega \leq \int f d\omega_a \leq \mu \int f d\omega , \tag{6}$$

for all $f \geq 0$ in $C(M)$. Thus,

$$\lambda\omega \leq \omega_a \leq \mu\omega , \tag{7}$$

so ω_a is absolutely continuous with respect to ω, and vice-versa. In particular, the notion of a set of ω_a measure 0, as well as the notion of a ω_a measurable set, are independent of the choice of a. We shall refer to a set of the first type as a set of <u>harmonic measure</u> 0, and a set of the second type as a harmonically measurable set. In this paper, the term measurable, when applied to a subset of M or a function on M, shall mean harmonically measurable, unless the contrary is stated. Similarly, a.e. shall mean a.e. with respect to harmonic measure.

From inequality 7, we also conclude that each of the spaces $L^p(d\omega_a)$, for $1 \leq p \leq \infty$, is independent of the choice of a, and different choices of a give equivalent L^p norms. (In this paper, $L^\infty(d\omega_a)$ shall denote the space of equivalence classes of complex valued essentially bounded measurable functions on M equal a.e.. The $L^\infty(d\omega_a)$ norm is the essential supremum norm.)

Since the mapping 2 is a lattice isomorphism, $C(M, \mathbb{R})$ is a complete vector lattice, and thus M is extremally disconnected [21, theorem 4.22.2]. We also have

Theorem 2: Each measurable, complex valued, essentially bounded function on M is equal a.e. to a unique function in $C(M)$.

Proof: First let f be a real, bounded, lower-semicontinuous function on M. Then $f = \sup A$, where the supremum is taken pointwise and

$$A = \{g \in C(M, \mathbb{R}): \ g \leq f\} .$$

The set A is directed. By the Harnack monotone convergence theorem, the pointwise supremum

$$h = \sup\{\phi^{-1}(g): \ g \in A\}$$

is harmonic, and in fact is equal to

$$\bigvee\{\phi^{-1}(g): \ g \in A\} .$$

Since $g \le \phi(h)$ for each $g \in A$ and $\phi(h)$ lies in $C(M, \mathbb{R})$, we have

$$f \le \phi(h) . \tag{8}$$

Thanks to equation 4, we also have

$$\int f d\omega = \int \phi(h) d\omega , \tag{9}$$

since the left hand side is simply

$$\sup\{\int g d\omega : \ g \in A\} = h(a_0) .$$

Now let f be an arbitrary bounded, real measurable function on M. Let

$$h = \bigwedge\{\phi^{-1}(g) : \ g \in B\} , \tag{10}$$

where

$$B = \{g \in C(M, \mathbb{R}) : \ f \le g\} .$$

Once again, the lattice infimum 10 is also a pointwise infimum. Thus,

$$\begin{aligned} \int f d\omega &= \inf\{\int g d\omega : \ g \in D\} \\ &= \inf\{\int g d\omega : \ g \in B\} \\ &= h(a_0) = \int \phi(h) d\omega , \end{aligned} \tag{11}$$

where the infima are taken pointwise, and

$D = \{g: g$ is real valued, bounded, lower-semicontinuous, and $f \leq g\}$.

The substitution of B for D in 11 is justified by 8 and 9. Set $s = \inf B$. Then s is upper-semicontinuous, and thus measurable, $f \leq s$, $\phi(h) \leq s$, and

$$\int s\,d\omega = \int \phi(h)\,d\omega \ .$$

Thus

$$\int |f-\phi(h)|\,d\omega \leq \int (s-f)\,d\omega + \int (s-\phi(h))\,d\omega = 0 \ ,$$

so $f = \phi(h)$ a.e..

It remains to show that $\phi(h)$ is the only element of $C(M)$ equal to f a.e.. Suppose $\phi(g) = f$ a.e. as well. Thanks to equation 4, for each $a \in R$,

$$|h(a)-g(a)| \leq \int |\phi(h)-\phi(g)|\,d\omega_a$$

$$\leq \int |\phi(h)-f|\,d\omega_a + \int |f-\phi(g)|\,d\omega_a = 0 \ ,$$

which shows $\phi(h) = \phi(g)$.

An alternate approach to uniqueness is to show that open subsets of M have positive harmonic measure. Let $U \subseteq M$ be open, and let $g \in C(M)$ be such that the support of g is contained in U, $g \geq 0$, and $g \not\equiv 0$. Then $\phi^{-1}(g) \geq 0$, and $\phi^{-1}(g) \not\equiv 0$. By the minimum principle,

$$0 < \phi^{-1}(g)\Big|_{a_0} = \int \phi(g)\,d\omega \leq ||g||_\infty \omega(U) \ .$$

Hayashi, by appealing to results of Dixmier, showed that M is extremally disconnected, and that each positive $f \in L^1(d\omega)$ is equal to a lower-semicontinous function a.e. [13]. In view of Hayashi's results, our interest in theorem 2 stems not from the innate interest of that theorem, but rather from its utility in the next chapter.

Let us turn now to the proof of theorem 1. First, we shall establish

Lemma 1: Let $f \in h^\infty(R)$. Then

$$\text{sp } f \subseteq \bigcap\{\text{Clf}(R\sim K)\colon\ K \subseteq R \text{ and compact}\},$$

where sp f denotes the spectrum of f .

Proof: Set

$$A = \bigcap\{\text{Clf}(R\sim K)\colon\ K \subseteq R \text{ and compact}\} .$$

Suppose $\lambda \notin A$. Let K be a compact subset of R such that $\lambda \notin \text{Clf}(R \sim K)$. Set $m = \inf\{|f(z)-\lambda|\colon\ z \in R \sim K\}$. Then $m > 0$. But for each $a \in R$,

$$m^2 \leq \lim_{S \uparrow J} \frac{1}{-2\pi} \int_{\partial S} |f-\lambda|^2 {*}dG_S(a,z) ,$$

so $m^2 \leq \langle f-\lambda, \overline{f-\lambda} \rangle$. Let $h = \langle f-\lambda, \overline{f-\lambda} \rangle$, and choose $k > 0$ so that $||kh||_\infty \leq 1$. Then $0 < km^2 \leq kh \leq 1$, so $||1-kh||_\infty < 1$. Consequently, h is invertible in $h^\infty(R)$. For invertible u in $h^\infty(R)$, let

$$u^{<-1>}$$

denote the inverse of u in $h^\infty(R)$. Then the inverse of $f - \lambda$ is

$$\langle \overline{f-\lambda}, h^{<-1>} \rangle ,$$

so $\lambda \notin$ sp f .

Inner functions on the unit disk give examples of functions f , for which the inclusion asserted by lemma 1 is strict. We shall now give the proof of theorem 1.

Proof: Let $f \in HB(R)$ be such that $\phi(f) \geq 0$. Then f has a square root v. Simply let $v = \phi^{-1}(\sqrt{\phi(f)})$, where the square root sign denotes the positive square root. Since

$$f = \langle v,v \rangle = \text{L. H. M. } |v|^2 ,$$

$f \geq 0$. Conversely, suppose $f \geq 0$. Thanks to lemma 1,

$$\text{Range } \phi(f) = \text{sp } f \subseteq \text{ClRange } f \subseteq \mathbb{R}^+ ,$$

where $\mathbb{R}^+$ denotes the non-negative reals. Thus $\phi(f) \geq 0$.

3.5. Throughout the rest of this chapter and the next, $1 \leq p \leq \infty$, $1 \leq q \leq \infty$, and $1/p + 1/q = 1$. As usual, if $p = \infty$, $q = 1$ and vice-versa. For $1 \leq p < \infty$, we shall need the formula,

$$\text{(1)} \qquad \text{L. H. M. } |f|^p \Big|_a = \lim_{S \in J} \frac{1}{-2\pi} \int_{\partial S} |f|^p {*dG_S(a,z)} ,$$

for each $f \in h^p(R)$.

The product $\langle \ , \ \rangle$ may be extended from $h^\infty(R) \times h^\infty(R)$ to $h^p(R) \times h^q(R)$. First, we shall carry out this extension for non-negative functions. Suppose f and g are non-negative functions in $h^p(R)$ and $h^q(R)$ respectively. Consider the net,

$$\text{(2)} \qquad \langle u,v \rangle , \quad 0 \leq u \leq f , \quad 0 \leq v \leq g , \quad u, v \in h^\infty(R) .$$

For $1 < p < \infty$, an application of Hölder inequality and 1 shows that

$$< u,v > \leq (\text{L. H. M. } f^p)^{1/p} (\text{L. H. M. } g^q)^{1/q} ,$$

for each element $< u,v >$ of the net 2. Similarly, if $p = 1$,

$$< u,v > \leq ||g||_\infty f ,$$

for each $< u,v >$ in the net 2. In either case, the net 2 is bounded above, and is monotone non-decreasing due to the monotoneity of $< \ , \ >$ on the product of the positive cone of HB(R) with itself. By the Harnack monotone convergence theorem, the pointwise supremum of the net 2 is harmonic. This justifies

Definition 1: Let $f \in h^p(R)$, $g \in h^q(R)$, and suppose $f \geq 0$ and $g \geq 0$. Then

$$< f,g > = \sup\{< u,v >: \ 0 \leq u \leq f, \ \ 0 \leq v \leq g, \ \ u, v \in HB(R)\} ,$$

where the supremum is taken pointwise.

Thanks to non-negativity, $< f,g > \in HP(R)$. Since $< f,g >$ is also quasi-bounded, $< f,g > \in h^1(R)$. It is clear that $< f,g >$ is in fact the HP(R) supremum of the net 2, and that the product $< \ , \ >$ is monotone increasing on the cartesian product of the positive functions in $h^p(R)$ with the positive functions in $h^q(R)$. The bilinearity of $< \ , \ >$ may be established with the help of the following lemma:

Lemma 1: Let f and g be as in definition 1 . Then,

$$\lim_{S \in J} \frac{1}{-2\pi} \int_{\partial S} |< f,g > - fg| *dG_S(a,z) = 0 .$$

Suppose that $1 \leq p < \infty$ and that f is a non-negative function in

$h^p(R)$. Suppose $0 \leq g \leq f$ (which implies $g \in h^p(R)$). From the elementary inequality

(3) $$(x-y)^p \leq x^p - y^p ,$$

for $0 \leq y \leq x$, we obtain

(4) $$\text{L. H. M. } (f-g)^p \leq \text{L. H. M. } f^p - \text{L. H. M. } g^p ,$$

thanks to 1.

Now suppose u_α , $\alpha \in A$, is a positive, monotone non-decreasing net in $h^p(R)$ converging pointwise to f . Since

$$\text{L. H. M. } u_\alpha^p \leq \text{L. H. M. } f^p$$

for each $\alpha \in A$, the net L. H. M. u_α^p , $\alpha \in A$, converges uniformly on compact subsets of R to a function $h \in h^1(R)$, as an application of the Harnack monotone convergence theorem shows. Clearly $h \leq$ L. H. M. f^p . Conversely, $u_\alpha^p \leq h$ for each $\alpha \in A$, so $f^p \leq h$. Thus L. H. M. $f^p \leq h$. We have now established

$$\lim_{\alpha \in A} \text{L. H. M. } u_\alpha^p = \text{L. H. M. } f^p .$$

From 4, we obtain

(5) $$\text{L. H. M. } (f-u_\alpha)^p \to 0 , \quad \alpha \in A .$$

In particular, $u_\alpha \to f$ in $h^p(R)$ norm. As a first consequence, this shows (if there were any doubt) that $h^\infty(R)$ is dense in $h^p(R)$. We may now begin the proof of lemma 1 .

Proof: Let $u_n = f \wedge n$ and $v_n = g \wedge n$, $n = 1,2,\cdots$. Thanks to non-negativity, u_n and $v_n \in h^\infty(R)$ for each n . Because f and g are quasi-bounded, $u_n \nearrow f$ and $v_n \nearrow g$ as $n \to \infty$. Furthermore, $< u_n,v_n > \nearrow < f,g >$ as $n \to \infty$ because the directed set $\{(u_n,v_n), n = 1,2,\cdots\}$ is cofinal in the index set of the net 2. Then for $S \in J$ and $n = 1,2,\cdots$,

(6) $$\frac{1}{-2\pi}\int_{\partial S} |< f,g > - fg| *dG_S(a_0,z) \le B_n + C_n + D_n ,$$

where

$$B_n = \frac{1}{-2\pi}\int_{\partial S} (< f,g > - < u_n,v_n >)*dG_S(a_0,z) ,$$

$$C_n = \frac{1}{-2\pi}\int_{\partial S} |< u_n,v_n > - u_nv_n| *dG_S(a_0,z) ,$$

$$D_n = \frac{1}{-2\pi}\int_{\partial S} (fg-u_nv_n)*dG_S(a_0,z) .$$

First suppose that $1 < p < \infty$. Note that

$$B_n = < f,g >\Big|_{a_0} - < u_n,v_n >\Big|_{a_0}$$

is independent of S , and clearly $\lim_{n\to\infty} B_n = 0$. Also,

(7) $$D_n \le ||f-u_n||^p||g||^q + ||f||^p||g-v_n||^q ,$$

where the norms raised to the exponent p are $h^p(R)$ norms, and the norms raised to the exponent q are $h^q(R)$ norms. To establish 7, consider the chain of inequalities,

$$D_n \leq \frac{1}{-2\pi}\int_{\partial S}((f-u_n)g + u_n(g-v_n))*dG_S(a_0,z)$$

$$\leq ||f-u_n||_S^p||g||_S^q + ||u_n||_S^p||g-v_n||_S^q \quad ,$$

where the norms on the right are $h^p(S)$ and $h^q(S)$ norms. The last member of this chain is dominated by the right hand side of 7 because the $h^p(S)$ norms (respectively $h^q(S)$ norms) converge monotonically upward to the $h^p(R)$ (respectively $h^q(R)$) norms. Note that the last line of 7 is independent of S . Thanks to 5, $\lim_{n\to\infty} D_n = 0$.

Let $\varepsilon > 0$. Choose n so large that $B_n < \varepsilon$ and $D_n < \varepsilon$. Using equation 3.2.8, choose $S_0 \in J$ so large that $C_n < \varepsilon$ for all $S \supseteq S_0$ contained in J . Applying 6, we obtain

$$\frac{1}{-2\pi}\int_{\partial S}|< f,g > - fg|*dG_S(a_0,z) < 3\varepsilon$$

for all $S \supseteq S_0$ contained in J , which establishes lemma 1 for $a = a_0$. Changing the base point if necessary, we obtain lemma 1 for $1 < p < \infty$ and for all $a \in R$. The proof for $p = 1$ or ∞ is similar.

In particular, for each $a \in R$,

$$\lim_{S\in J}\frac{1}{-2\pi}\int_{\partial S} fg*dG_S(a,z) = \lim_{S\in J}\frac{1}{-2\pi}\int_{\partial S}< f,g >*dG_S(a,z) = < f,g >\Big|_a \quad ,$$

if $f \in h^p(R)$, $g \in h^q(R)$, and $f, \ g \geq 0$. This justifies

Definition 2: Let $f \in h^p(R)$ and $g \in h^q(R)$. Both f and g may be complex valued. Then

$$< f,g >\Big|_a = \lim_{S\in J}\frac{1}{-2\pi}\int_{\partial S} fg*dG_S(a,z) \ .$$

By splitting f and g into linear combinations of positive functions in $h^p(R)$ and $h^q(R)$ respectively, we see that the limit defining $< f,g >$ exists, and that $< f,g > \in h^1(R)$. (To carry out the process of splitting f and g up, we require the obvious fact that for real valued $u \in h^p(R)$, $u \wedge 0$ and $u \vee 0$ both lie in $h^p(R)$.) Clearly $< \ , \ >$ is bilinear and commutative. From lemma 1, we immediately obtain the first assertion of

Theorem 1: Let $f \in h^p(R)$ and $g \in h^q(R)$. Then the following are true:

A. $$\lim_{S \in J} \frac{1}{-2\pi} \int_{\partial S} |< f,g > - fg| *dG_S(a,z) = 0, \text{ for each } a \in R .$$

B. If $1 < p < \infty$, then

$$\text{L. H. M. } |< f,g >| \leq (\text{L. H. M. } |f|^p)^{1/p} (\text{L. H. M. } |g|^q)^{1/q} .$$

C. If $p = \infty$, then

$$\text{L. H. M. } |< f,g >| \leq ||f||_\infty \text{ L. H. M. } |g|$$

Proof: Consider, for example, assertion B. Assertion B follows immediately from equation 1, definition 2, and the Hölder inequality.

Corollary 1: The bilinear map $< \ , \ >: h^p(R) \times h^q(R) \to h^1(R)$ is continuous.

Proof: Evaluating both sides of B and C at a_0, we obtain

$$(8) \qquad ||< f,g >||_1 \leq ||f||_p ||g||_q ,$$

where $|| \ ||_p$ and $|| \ ||_q$ denote the $h^p(R)$ and $h^q(R)$ norms respectively.

In the next chapter, we shall obtain two results closely related to theorem 1. We shall show that $h^p(R)$ is an involutive $h^\infty(R)$ Banach module, when endowed with the $h^p(R)$ norm, with $< \; , \; >$ as multiplication, and with complex conjugation as involution. We shall also show that $h^p(R)$ is an involutive, locally convex $(h^\infty(R), \beta)$ module, when $h^p(R)$ is endowed with the weak topology (with the weak * topology if $p = \infty$). We shall make these assertions precise later.

For f and $g \in h^2(R)$ and real, we have

$$(9) \qquad < f,g > = \frac{1}{4}(\text{L. H. M. } (f+g)^2 - \text{L. H. M. } (f-g)^2) \ ,$$

by the same argument used to show the existence of the limit 3.2.2. Further, we have

Theorem 2: Let f and $g \in h^2(R)$. Then

$$\frac{1}{-2\pi} \int_{\partial S} |< f,g > - fg| * dG_S(a_0, z)$$

$$\leq (||f||^2 - ||f||_S^2)^{1/2} (||g||^2 - ||g||_S^2)^{1/2} \ ,$$

for each $S \in J$.

The proof and notation are as in lemma 3.2.2. As explained in the remarks following lemma 3.2.2, a similar inequality does not seem to be possible for $p \neq 2$.

Equation 9 is rather interesting. The Banach space $h^2(R)$ becomes a Hilbert space when endowed with the sesquilinear form

$$(f,g) \longrightarrow < f,\bar{g} >\Big|_{a_0}$$

In fact, $< f,\bar{f} >\Big|_{a_0}$ is simply the $h^2(R)$ norm of f , as explained in equation

3.2.11. Thus equation 9 is a sort of generalized polarization identity giving the inner product in terms of the norm.

3.6. There are two simple interpretations of the product $< \, , \, >$ which we shall give in this section. First though, we shall need a uniqueness lemma: Let f and g be real valued functions in $h^p(R)$ and $h^q(R)$ respectively. Suppose that u_α and v_α, $\alpha \in A$, are any pair of monotone nets in $h^p(R)$ and $h^q(R)$ respectively, such that $u_\alpha \nearrow f$ and $v_\alpha \nearrow g$ pointwise. From equation 3.5.5 and corollary 3.5.1.1, we conclude

$$(1) \qquad < u_\alpha, v_\alpha > \nearrow < f,g >$$

both pointwise and in $h^1(R)$ norm. If $p = 1$ or ∞, we must apply inequality 3.5.8 rather than corollary 3.5.1.1.

Let

$$[\, , \,]: \; h^p(R) \times h^q(R) \longrightarrow h^1(R)$$

be another bilinear map. Suppose $[\, , \,]$ agrees with $< \, , \, >$ on all pairs of positive functions in $h^\infty(R) \times h^\infty(R)$. From 1 and bilinearity, we immediately obtain the following uniqueness lemma:

<u>Lemma</u> 1: Suppose for each pair of positive functions f and g in $h^p(R)$ and $h^q(R)$ respectively, and for each pair of monotone non-decreasing sequences f_n and g_n, $n = 1, 2, \ldots$, in $h^\infty(R)$, such that $f_n \nearrow f$ and $g_n \nearrow g$ pointwise as $n \to \infty$, we have

$$\lim_{n \to \infty} [f_n, g_n] = [f,g] .$$

Then $[\, , \,] = < \, , \, >$ on all of $h^p(R) \times h^q(R)$.

In particular, if $[\, , \,]$ is continuous on $h^p(R) \times h^q(R)$, we have $[\, , \,] = < \, , \, >$ on $h^p(R) \times h^q(R)$. As an application of this remark, we shall give a simple interpretation of $< \, , \, >$ in terms of the Martin boundary.

We return to the notation of section 3.3. As in the derivation of equation 3.3.2, we obtain for $1 \leq p < \infty$,

(2) $$\text{L. H. M. } |f|^p\Big|_a = \int_{\Delta_1} |\hat{f}|^p d\omega_a \ ,$$

for each $a \in R$ and $f \in h^p(R)$. Also,

(3) $$||f||_\infty = ||\hat{f}||_\infty \ ,$$

for each $f \in h^\infty(R)$. The norm on the right in 3 is the $L^\infty(d\omega_{a_0})$ norm of f . From equations 2.3, 3.3.1, and from the uniqueness of the representing measure for $hP(R)$ functions [5, p. 138], we see that the mapping

(4) $$v \longrightarrow \int_{\Delta_1} v d\omega_a \ , \quad a \in R \ ,$$

is an isometric isomorphism from $L^p(d\omega_{a_0})$ onto $h^p(R)$, and the mapping

(5) $$f \longrightarrow \hat{f} \ , \quad f \in h^p(R) \ ,$$

is the inverse of the map 4.

For each $a \in R$, set

(6) $$[\ f,g\ ]\Big|_a = \int_{\Delta_1} \hat{f}\hat{g} d\omega_a \ ,$$

for all $(f,g) \in h^p(R) \times h^q(R)$. An application of 4, 5, and the Hölder inequality shows both that the range of $[\ ,\ ]$ is contained in $h^1(R)$ for each $1 \leq p \leq \infty$, and that the map

(7) $$[\ ,\ ]:\ h^p(R) \times h^q(R) \longrightarrow h^1(R)$$

is continuous. Thanks to equation 3.3.3 and the remark following lemma 1, we have $[\ ,\ ] = <\ ,\ >$ on all of $h^p(R) \times h^q(R)$. Thus for each $a \in R$,

(8) $$< f,g > \Big|_a = \int_{\Delta_1} \hat{f}\hat{g}\, d\omega_a \ ,$$

for all $(f,g) \in h^p(R) \times h^q(R)$.

If $R = \Delta(0,1)$, the preceding remarks may be duplicated almost word for word, and we obtain for all $(f,g) \in h^p(R) \times h^q(R)$,

(9) $$< f,g > \Big|_a = \int_0^{2\pi} f^* g^* K(a,e^{i\theta}) \frac{d\theta}{2\pi} \ .$$

If R is an arbitrary hyperbolic Riemann surface, we may apply lemma 1 and equation 3.3.6 to obtain,

(10) $$< f,g > \circ\ \psi = < f\circ\psi, g\circ\psi > \ ,$$

for all $(f,g) \in h^p(R) \times h^q(R)$.

Either 8, or 9 and 10 together, give us a reasonably simple interpretation of $<\ ,\ >$. For the reasons explained in section 3.3 though, we shall not base our subsequent development of the product $<\ ,\ >$ on either of these interpretations. We shall use neither the results of this section, except for lemma 1, nor of section 3.3 elsewhere in the body of this paper.

Chapter 4

An Operational Calculus, Duality

4.1. We return to the notation introduced in section 3.4. In particular, throughout the rest of this paper, M shall be the Hayashi boundary of R, that is, M shall be the maximal ideal space of the B^* algebra $h^\infty(R)$. We shall let

$$\phi\colon\ h^\infty(R) \longrightarrow C(M)$$

denote the Gelfand transform. (We shall change the definition of ϕ slightly from section 4.2. on.)

The results of chapter 3 would be of rather modest interest, if it were not for the fact that the isomorphism ϕ can be extended to an isomorphism from $h^p(R)$ <u>onto</u> $L^p(d\omega)$, for each $1 \le p \le \infty$. We shall prove this theorem in this chapter, along with several other useful results about the duals of $h^p(R)$ and $(h^\infty(R),\beta)$. We shall also develop representations of integrals of the general type

$$\int f d\nu = \left.\langle \phi(f), \int \phi(q_a) d\nu(a)\rangle\right|_{a_0} ,$$

where q_a is a reproducing kernel for $h^2(R)$. Most interesting of all, we shall obtain an extension of the Szëgo maximal theorem, which gives us a formula for the outer factor of an $H^1(R)$ function in terms of an integral over M.

All of these results follow from an operational calculus for $h^\infty(R)$, which we shall now develop. Let $P\colon\ A \longrightarrow \mathbb{C}$ be continuous, where

$A \subseteq C^n$. We shall let

$$(1) \qquad \langle P\rangle(h_1,\ldots,h_n) = \phi^{-1}(P \circ (\phi(h_1),\ldots,\phi(h_n))),$$

for each n-tuple $(h_1,\ldots,h_n) \in (h^\infty(R))^n$ with Range $(\phi(h_1,\ldots,\phi(h_n)) \subseteq A$. For $h \in h^\infty(R)$ and $s \in \mathbb{R}$,

$$(2) \qquad h^{\langle s\rangle} = \phi^{-1}((\phi(h))^s),$$

assuming $(\phi(h))^s$ is well defined. In particular, if s is not integral, $h^{\langle s\rangle}$ shall be defined only for positive h, and $h^{\langle s\rangle}$ shall be positive for positive h.

The fundamental theorem of our operational calculus is

Theorem 1: Let $A \subseteq \mathbb{C}^n$, and let $P\colon A \longrightarrow \mathbb{C}$ be continuous. Then for each $a \in R$,

$$\lim_{S \in J} \frac{1}{-2\pi} \int_{\partial S} |\langle P\rangle(h) - P \circ h| {*dG_S(a,z)} = 0 ,$$

where $h = (h_1,\cdots,h_n)$ is any element of $(h^\infty(R))^n$ such that

$$(3) \qquad \prod_{i=1}^{n} \mathrm{Cl}h_i(R\sim K) \subseteq A ,$$

for some compact $K \subseteq R$.

Proof: From theorem 3.2.1 and equation 3.2.8, we immediately see that the conclusion of theorem 1 holds if $P(z_1,\cdots,z_n)$ is a polynomial in $(z_1,\cdots,z_n)$ and $(\bar{z}_1,\cdots,\bar{z}_n)$. Suppose that P is continuous on A, and let h be an element of $(h^\infty(R))^n$ satisfying inclusion 3. Let B denote the left hand side of that inclusion. Note that B is compact. By lemma 3.4.1, $\prod_{i=1}^{n} \phi(h_i)(M) \subseteq B$, so $< P >(h)$ is well defined. The polynomials in $(z_1, \cdots, z_n)$ and $(\bar{z}_1, \cdots, \bar{z}_n)$ are a subalgebra of $C(B)$ containing conjugates and separating the points of B. An application of the Stone-Weierstrass theorem shows that this polynomial subalgebra is dense in $C(B)$. Here, $C(B)$ is the Banach algebra of continuous, complex valued functions on B. Let $\varepsilon > 0$, and let Q be such a polynomial satisfying $||P-Q||_\infty < \varepsilon$, where the supremum norm is taken in $C(B)$. Let $a \in R$, and let $S_0 \in J$ be so large that $K \subseteq S_0$, $a \in S_0$, and

$$\frac{1}{-2\pi} \int_{\partial S} |< Q >(h) - Q \circ h| *dG_S(a,z) < \varepsilon ,$$

for all $S_0 \subseteq S \in J$. For all such S, $|P\circ h - Q\circ h| < \varepsilon$ on ∂S. In addition,

$$||< P >(h) - < Q >(h)||_\infty$$

$$= ||\phi^{-1}(P\circ\phi(h) - Q\circ\phi(h))||_\infty$$

$$= ||P\circ\phi(h) - Q\circ\phi(h)||_\infty < \varepsilon ,$$

since ϕ is an isometry. Here $\phi(h) = (\phi(h_1), \cdots \phi(h_n))$. Thus for $S_0 \subseteq S \in J$,

$$\frac{1}{-2\pi} \int_{\partial S} |< P >(h) - P \circ h| *dG_S(a,z)$$

$$\leq \frac{1}{-2\pi} \int_{\partial S} |< P >(h) - < Q >(h)| *dG_S(a,z)$$

$$+ \frac{1}{-2\pi} \int_{\partial S} |< Q >(h) - Q \circ h| *dG_S(a,z)$$

$$+ \frac{1}{-2\pi} \int_{\partial S} |Q\circ h - P\circ h| *dG_S(a,z) < 3\varepsilon ,$$

which establishes the theorem.

From theorem 1 we immediately conclude

(4) $$\langle P \rangle(h)\Big|_a = \lim_{S\in J} \frac{1}{-2\pi}\int_{\partial S} P\circ h * dG_S(a,z) ,$$

for each $a \in R$ and each pair of functions P and h satisfying the hypotheses of theorem 1. In particular, for each $1 \le p < \infty$,

$$\phi^{-1}(|\phi(h)|^p)\Big|_a = \lim_{S\in J} \frac{1}{-2\pi}\int_{\partial S} |h|^p * dG_S(a,z) ,$$

for each $h \in h^\infty(R)$. Thanks to equations 3.4.4 and 3.5.1, for each $1 \le p < \infty$,

(5) $$\int |\phi(h)|^p d\omega_a = \text{L. H. M. } |f|^p\Big|_a ,$$

for each $h \in h^\infty(R)$.

4.2. Equation 4.1.5 may be used to extend ϕ linearly and isometrically to $h^p(R)$. First let $1 \le p < \infty$, and consider the map between normed linear spaces,

(1) $$\phi\colon\ (h^\infty(R), \|\ \|_p) \longrightarrow (L^\infty(d\omega), \|\ \|_p) ,$$

where the norms are the $h^p(R)$ and $L^p(d\omega)$ norms respectively. Here, ϕ is defined by $\phi(f)$ = the equivalence class in $L^\infty(d\omega)$ containing the Gelfand transform of f, for each $f \in h^\infty(R)$. Appealing to theorem 3.4.2, we see that the map 1 is an onto isomorphism. Equation 4.1.5 implies that ϕ is an isometry, that is $\|f\|_p = \|\phi(f)\|_p$ for each $f \in h^\infty(R)$. Since the $h^p(R)$ norm completion of $h^\infty(R)$ is $h^p(R)$, and the $L^p(d\omega)$ norm completion of $L^\infty(d\omega)$ is $L^p(d\omega)$, the map 1 may be extended uniquely by continuity to a linear isometry

(2) $$\phi\colon\ h^p(R) \xrightarrow{\text{onto}} L^p(d\omega) .$$

Clearly,

(3) $$\phi^{-1}(f)\Big|_a = \int f d\omega_a ,$$

since both sides are continuous in $L^p(d\omega)$ norm, and agree on $L^\infty(d\omega)$. Also, since ϕ is an isometry,

$$\text{L. H. M. } |h|^p\Big|_{a_0} = \int |\phi(h)|^p d\omega_{a_0}$$

By changing the base point a_0 , we see that

(4) $$\text{L. H. M. } |h|^p\Big|_a = \int |\phi(h)|^p d\omega_a ,$$

for each $h \in h^p(R)$ and $a \in R$.

Because the map 1 is an isometry for each p , the map

(5) $$\phi\colon\ h^\infty(R) \xrightarrow{\text{onto}} L^\infty(d\omega) ,$$

is also an isometry. Here, as usual, $h^\infty(R)$ is endowed with the $h^\infty(R)$ norm, and $L^\infty(d\omega)$ is endowed with the essential supremum norm. Temporarily, let $|| \ ||_e$ be the essential supremum norm, and let $|| \ ||_\infty$ be the supremum norm on R or M . It will be clear from the context which is meant. Let $h \in h^\infty(R)$, and let g be the Gelfand transform of h . Because $g \in \phi(h)$, $||\phi(h)||_e \le ||g||_\infty = ||h||_\infty$. Conversely, $||\phi(h)||_e = \sup\{||\phi(h)||_p\colon\ 1 \le p < \infty\}$ $= \sup\{||h||_p\colon\ 1 \le p < \infty\}$. However, for each $S \in J$ and $1 \le p < \infty$,

$$\left(\int_{\partial S} |h|^p {*dG_S(a_0,z)}\right)^{1/p} \le ||h||_p ,$$

and

$$\sup\left\{\left(\int_{\partial S} |h|^p {*dG_S(a_0,z)}\right)^{1/p}\colon\ 1 \le p < \infty\right\}$$

$$= \sup\{|h(z)|: z \in \partial S\} = \sup\{|h(z)|: z \in S\} ,$$

by standard L^p theory and the maximum principle. Thus $||h||_\infty \leq ||\phi(h)||_e$.

Throughout the remainder of this paper, ϕ shall denote either the map 2 or the map 5. We have already shown the first part of

Theorem 1: The mapping ϕ is a linear onto isometry which preserves both involution and order.

Proof: Only the parts about involution and order need proving. Because the Gelfand transform preserves involution, $\phi(\overline{f}) = \overline{\phi(f)}$ a.e. for all $f \in h^p(R)$. Suppose that $f \in h^p(R)$, and that $\phi(f) \geq 0$ a.e.. Thanks to equation 3, $f \geq 0$. Conversely, suppose $f \geq 0$. First let $p = \infty$. Because the Gelfand transform preserves order, $\phi(f) \geq 0$ a.e.. Now suppose $1 \leq p < \infty$. Then the sequence $f \wedge n$, $n = 1,2,\cdots$, is non-negative, lies in $h^\infty(R)$, and converges upward in $h^p(R)$ norm to f . Thanks to the continuity of ϕ ,

$$0 \leq \lim_{n\to\infty} \phi(f \wedge n) = \phi(f) \quad \text{a.e.} \quad .$$

In particular, because ϕ is onto, ϕ is a lattice isomorphism between the lattices of real valued functions in $h^p(R)$ and the real valued functions in $L^p(d\omega)$. This gives us a particularly simple interpretation for the lattice supremum in HP(R): For f and g real valued and in $h^1(R)$,

$$(6) \qquad f \vee g\Big|_a = \int \max(\phi(f), \phi(g)) d\omega_a ,$$

for each $a \in R$. We also have

Theorem 2: The map ϕ is also multiplicative, that is,

$$\phi(< f,g >) = \phi(f)\phi(g) ,$$

for all $f \in h^p(R)$ and $g \in h^q(R)$.

Proof: Temporarily let

$$[\, f,g \,]\Big|_a = \int \phi(f)\phi(g)d\omega_a \,,$$

for $(f,g) \in h^p(R) \times h^q(R)$. The bilinear map

$$[\ ,\ ]:\ h^p(R) \times h^q(R) \longrightarrow h^1(R)\,,$$

is continuous, and agrees with $<\ ,\ >$ on $h^\infty(R) \times h^\infty(R)$. By lemma 3.6.1, $[\ ,\ ] = <\ ,\ >$ on $h^p(R) \times h^q(R)$.

Let A be a Banach algebra, and let B be a Banach space. Let

$$\cdot :\ A \times B \longrightarrow B$$

be a bilinear map such that $(ab) \cdot c = a \cdot (b \cdot c)$ for all $(a,b,c) \in A \times A \times B$. The pair $(B,\cdot)$ is a (left) A Banach module if

(7) $$||a \cdot b|| \le ||a||\ ||b||\,,$$

for all $(a,b) \in A \times B$. If A is a commutative Banach algebra with involution $*$, and if B is a Banach space with involution, also denoted by $*$, then $(B,\cdot)$ is an involutive A Banach module if $(a \cdot b)^* = a^* \cdot b^*$. Suppose A is a locally convex topological algebra (we require joint continuity of the product), and B is a locally convex space. The notions of a (left) A locally convex module and an involutive A locally convex module are defined similarly, except that inequality 7 is replaced by the requirement that the map $\cdot$ be separately continuous.

Theorem 3:

A. $h^p(R)$ is an involutive $h^\infty(R)$ Banach module under $<\ ,\ >$ as multiplication and complex conjugation as involution. The map ϕ is an involutive Banach module isomorphism.

B. The map $\phi: h^\infty(R) \xrightarrow{\text{onto}} L^\infty(d\omega)$ is a B^* algebra isomorphism.

Proof: Assertion B follows from A, and is stated only for emphasis. To prove A, we must show, for each pair $(f,g) \in h^\infty(R) \times h^p(R)$, that $< f,g > \in h^p(R)$, that $\|< f,g >\|_p \le \|f\|_\infty \|g\|_p$, and that $< \bar{f},\bar{g} > = \overline{< f,g >}$. In addition for all (f,g,h) lying in $(h^\infty(R) \times h^\infty(R) \times h^p(R))$, we must show $<< f,g >,h > = < f,< g,h >>$. All this is clear from theorems 1 and 2, and from the corresponding results for $L^p(d\omega)$ as an $L^\infty(d\omega)$ module under pointwise a.e. multiplication.

The same argument shows

$$< f,< g,h >> = << f,g >,h > , \tag{8}$$

for all triples (f,g,h) for which two coordinates lie in $h^p(R)$ and $h^q(R)$ respectively, and the remaining coordinate lies in $h^\infty(R)$. For simplicity, we shall sometimes write

$$< f,g,h > \tag{9}$$

instead of $< f,< g,h >>$ or $<< f,g >,h >$. Applying inequality 3.2.9, we obtain

Theorem 4: Let $f \in h^\infty(R)$, $g \in h^p(R)$, and $h \in h^q(R)$. Then for each $a \in R$,

$$< f,g,h > \Big|_a = \lim_{S \in J} \frac{1}{-2\pi} \int_{\partial S} fgh{*}dG_S(a,z) ,$$

and,

$$\lim_{S \in J} \frac{1}{-2\pi} \int_{\partial S} |< f,g,h > - fgh|{*}dG_S(a,z) = 0 .$$

4.3. We return to the operational calculus developed in section 4.1. Let $P: A \longrightarrow \mathbb{C}$ be continuous, where $A \subseteq \mathbb{C}^n$. Thanks to equations 3.4.4 and 4.2.3, even though we have changed the definition of ϕ on sets of measure 0,

(1) $$\langle P \rangle(h_1,\cdots,h_n) = \phi^{-1}(P\circ(\phi(h_1),\cdots,\phi(h_n))) ,$$

for each n-tuple $(h_1,\cdots,h_n)$ satisfying the conditions following equation 4.1.1. Thus, theorem 4.1.1 provides the beginnings of an operational calculus relative to the isomorphism

$$\phi:\ h^\infty(R) \longrightarrow L^\infty(d\omega) .$$

This operational calculus may be extended much further. We have already observed in equation 4.2.4 that the formula

$$\phi^{-1}(|\phi(h)|^p)\Big|_a = \lim_{S\in J} \frac{1}{-2\pi}\int_{\partial S} |h|^p {*dG_S(a,z)} ,$$

extends to all $h \in h^p(R)$. We shall content ourselves with the following two extensions. The first is a trivial extension of theorem 4.2.4.

Theorem 1: Let $P(z_1,\cdots,z_n)$ be a polynomial in $(z_1,\cdots,z_n)$ and $(\bar{z}_1,\cdots,\bar{z}_n)$. Let $h_1,\cdots,h_n$ be harmonic on R, suppose h_1 belongs to $h^p(R)$, h_2 belongs to $h^q(R)$, and all the rest of the h_i's belong to $h^\infty(R)$. In addition, suppose that z_1, $\bar{z}_1$, z_2, and $\bar{z}_2$ occur at most with degree 1 in each term of $P(z_1,\cdots,z_n)$. Then for each $a \in R$,

$$\langle P \rangle(h)\Big|_a = \lim_{S\in J} \frac{1}{-2\pi}\int_{\partial S} P\circ h {*dG_S(a,z)} ,$$

and

$$\lim_{S\in J} \frac{1}{-2\pi}\int_{\partial S} |\langle P \rangle(h) - P\circ h| {*dG_S(a,z)} = 0 .$$

Here, $h = (h_1,\cdots,h_n)$.

The second extension is a generalization of equation 4.2.4. Let α be a function on $[-\infty,+\infty]$ satisfying the de la Vallée Poussin condition (c.f. section 2.5). Then

Theorem 2: Let $h \in Q(R)$, and suppose $\alpha \circ h$ admits a harmonic majorant. Then

$$\alpha \circ \phi(h) = \phi(\text{L. H. M.}\,\alpha \circ h) \quad \text{a.e.} \quad .$$

Proof: First suppose $h \in HB(R)$. From equation 4.3.1, we conclude

$$(2) \qquad \lim_{S \in J} \int \alpha \circ h * dG_S(a,z) = \phi^{-1}(\alpha \circ \phi(h))\Big|_a$$

Thanks to the subharmonicity of $\alpha \circ h$, the left hand side of (2) is simply L. H. M. $\alpha \circ h$.

For $-\infty < x_1 < x_2 < +\infty$, let $m(x_1,x_2)$ denote the slope of the chord from $(x_1,\alpha(x_1))$ to $(x_2,\alpha(x_2))$. For each $s \in \mathbb{R}$, the convexity of α implies that the function $x \to m(x,s)$ is non-decreasing on $(-\infty,s)$. Set $k(s) = \lim_{x \to s-} m(x,s)$. Because α is convex and non-decreasing,

$$(3) \qquad 0 \le \alpha(x) - \alpha(x_1) \le k(s)(x - x_1) \; ,$$

for all $-\infty < x_1 < x < s$.

Let $h \in Q(R)$, and suppose h is bounded above. Let h_n, $n = 1,2,\cdots$ be a non-increasing sequence in $HB(R)$ such that $h_n \downarrow h$ pointwise. Clearly $\alpha \circ h \le \alpha \circ h_n$ for each n, so $\alpha \circ h$ admits a harmonic majorant. Let $s \in \mathbb{R}$ be so large that $h \le s$ and $h_n \le s$ for $n = 1,2,\cdots$. Applying equation 3, we obtain for each $a \in R$,

$$\begin{aligned}
0 &\le (\text{L. H. M. } \alpha(h_n) - \text{L. H. M. } (\alpha \circ h))\Big|_a \\
&= \lim_{S \in J} \int_{\partial S} (\alpha \circ h_n - \alpha \circ h) * dG_S(a,z) \\
&\le \lim_{S \in J} k(s) \int_{\partial S} (h_n - h) * dG_S(a,z) \\
&= k(s)(h_n(a) - h(a)) \; .
\end{aligned}$$

In particular, L. H. M. $\alpha \circ h_n \downarrow$ L. H. M. $\alpha \circ h$ in $h^1(R)$ norm. Because the map 4.2.2 is an isometry for $p = 1$, we obtain

(4) $$\phi(\text{L. H. M. } \alpha o h) = \lim_{n\to\infty} \phi(\text{L. H. M. } \alpha o h_n)$$

$$= \lim_{n\to\infty} \alpha o \phi(h_n) = \alpha o \phi(h) \quad \text{a.e.} \quad .$$

Now let h be an arbitrary element of $Q(R)$ such that $\alpha \circ h$ admits a harmonic majorant. Let h_n, $n = 1,2,\cdots$ be a non-decreasing sequence in $Q(R)$ such that each h_n is bounded above, and $h_n \nearrow h$ pointwise as $n \to \infty$. Thanks to equation 3.5.5, $h_n \to h$ in $h^1(R)$ norm. As in the derivation of equation 4,

$$\phi(\text{L. H. M. } \alpha o h) = \alpha o \phi(h) \quad \text{a.e.} \quad .$$

With the aid of theorem 2 and Heins' extended theorem of Szegö-Solomentsev (theorem 2.5.1), we may obtain a generalization of the Szegö maximal theorem. Our proof will be patterned after Heins' derivation of the classical Szegö maximal theorem [17, p. 27]. Let $\alpha(x) = \exp x$ for $x \in R$, and extend α by continuity to $[-\infty,+\infty]$. Let $f \in H^1(R)$, and let $h \in Q(R)$ be such that $\exp h$ is the outer factor of $|f|$. By theorem 2.5.3, L. H. M. $\exp h =$ L. H. M. $|f|$. Applying theorem 2 and equation 4.2.4, we obtain

(5) $$\exp \phi(h) = \phi(\text{L. H. M. } |f|) = |\phi(f)| \quad \text{a.e.} \quad .$$

Thus $\ln|\phi(f)| = \phi(h)$ a.e., and consequently both $\ln|\phi(f)|$ and $\min(\ln|\phi(f)|,0)$ are integrable $d\omega$. We have now given the proof of most of our generalization of the Szegö maximal theorem.

<u>Theorem</u> 3: Let $f \in H^1(R)$, and suppose $f \not\equiv 0$. Then $\phi(f)$ differs from 0 and ∞ a.e.. The function $\ln|\phi(f)|$ is integrable $d\omega$, and the outer factor of f is

$$F(a) = \exp\left(\int \ln|\phi(f)|\,d\omega_a\right), \quad a \in R .$$

Furthermore, among the class of all l.a.m.'s H admitting a harmonic majorant such that $\phi(\text{L. H. M. } H) = |\phi(f)|$, F is maximal.

Proof: We need only show that given an l. a. m. H admitting a harmonic majorant such that $\phi(\text{L. H. M. } H) = |\phi(f)|$, we have $H(a) \leq F(a)$, for all $a \in R$. By the same argument used to establish 5, F is the outer factor of any such l.a.m. H. Thanks to theorem 2.5.3 again, the inner factor of H is bounded above by 1. This establishes the theorem.

4.4. With the aid of the onto isometric isomorphism ϕ, we may easily attack the problem of duality for $h^p(R)$ spaces. Throughout this paper, $[\ ,\ ]$ shall denote the bilinear form $\langle\ ,\ \rangle\big|_{a_0}$. From $L^p - L^q$ duality and theorem 4.2.2, we immediately obtain

Theorem 1: Let $1 \leq p < \infty$. Then the $h^p(R)$ norm dual of $h^p(R)$ is $h^q(R)$, and the dual norm is the $h^q(R)$ norm, via the bilinear form $[\ ,\ ]$.

Theorem 1 is particularly useful for obtaining integral representation formulas of the type promised at the beginning of this chapter. The rest of this section will be devoted to such representations.

Let $a \in R$, and let λ and μ be the Harnack constants appearing in inequality 3.4.7. Let k_a be the Radon-Nikokym derivative

(1) $$k_a = \frac{d\omega_a}{d\omega} .$$

Since $\lambda\omega \leq \omega_a \leq \mu\omega$, we have

(2) $$\lambda \leq k_a \leq \mu \quad \text{a.e.} .$$

Thus $k_a \in L^\infty(d\omega)$. Set

(3) $$q_a = \phi^{-1}(k_a).$$

Clearly q_a is a positive element of $HB(R)$. In fact, since the measure ω has total mass 1,

(4) $$\lambda \leq q_a \leq \mu \quad ,$$

From the equation

(6) $$f(a) = \int \phi(f) k_a d\omega \ , \quad f \in h^1(R) \ ,$$

we obtain

(7) $$f(a) = [\ f, q_a\ ] \ .$$

for each $f \in h^1(R)$. Clearly, q_a is the unique element of $h^1(R)$ for which equation 7 holds for all $f \in h^\infty(R)$.

The function

$$(a,z) \longrightarrow q_a(z) = q(a,z) \ , \quad (a,z) \in R \times R \ ,$$

is symmetric by the usual argument for reproducing kernels. Thus $q(a,z)$ is harmonic in each variable. In addition, $q(a,z)$ is jointly continuous, and if $K \subseteq R$ is compact, the family q_a , $a \in K$, is uniformly bounded in $h^\infty(R)$ norm. The proof of this is subsumed in the proof of the following lemma. Here, and wherever necessary throughout the rest of this paper, k_a shall denote the continous function equal to the function given by equation 1 a.e..

Lemma 1: The function

$$(a,t) \longrightarrow k_a(t) = k(a,t) \ , \quad (a,t) \in R \times M \ ,$$

is jointly continuous on $R \times M$. Furthermore, if $K \subseteq R$ is compact, then the family in $C(M)$, k_a , $a \in K$, is uniformly bounded in supremum norm.

Proof: Let $K \subseteq R$ be compact, and let $0 < \lambda(K) \leq \mu(K) < \infty$ be the Harnack constants for K . As in the derivation of inequality 3.4.7 and inequality 2,

(8) $$\lambda(K)k_a \leq k_b \leq \mu(K)k_a ,$$

for all a and b lying in K. However, $\lambda(K)$ and $\mu(K) \to 1$ as $K \to \{a\}$. That is, given $a \in R$ and $\varepsilon > 0$, there exists a neighborhood U of a such that $|\lambda(K) - 1|$ and $|\mu(K) - 1| < \varepsilon$, for all compact K containing a such that $K \subseteq U$. Thus $k_b \to k_a$ in supremum norm as $b \to a$. Joint continuity now follows easily.

By choosing K such that $a_0 \in K$, and letting $a = a_0$ in inequality 8, we immediately obtain the assertion about uniform boundedness. In fact, we see that inequality 2 holds uniformly for $a \in K$, provided λ and μ are replaced by $\lambda(K)$ and $\mu(K)$ respectively.

Theorem 2: Let ν be a complex regular Baire measure on R of bounded total variation.

A. The function

(9) $$\int q_a d\nu(a)$$

lies in $h^1(R)$, and

(10) $$\int f d\nu = [\ f, \int q_a d\nu(a)\ ] ,$$

for all $f \in h^\infty(R)$.

B. If ν has compact support, then the function 9 lies in $h^\infty(R)$, and equation 10 holds for all $f \in h^1(R)$.

The integral 9 is defined pointwise, that is

(11) $$\int q_a d\nu(a)\Big|_z = \int q(a,z) d\nu(a) ,$$

for all $z \in R$. Similarly, we shall let

(12) $$\int k_a d\nu(a)\Big|_t = \int k(a,t) d\nu(a) ,$$

provided the right hand integral exists.

Proof: It suffices to assume ν is positive. First suppose ν has compact support K. Because $k(a,t)$ is uniformly continuous on the compact set $K \times M$, the function 12 is continuous on M. Applying Fubini's theorem and equation 3, we have

$$\int_K q(a,z)d\nu(a) = \int_M \int_K k(a,t)d\nu(a)d\omega_z(t) .$$

Thus

$$(13) \qquad \int q_a d\nu(a) = \phi^{-1}\left(\int_K k_a d\nu(a)\right) .$$

In particular, the function 9 lies in $h^\infty(R)$. For $f \in h^1(R)$, we have by applying Fubini's theorem again,

$$\begin{aligned}\int f d\nu(a) &= \int_M \int_K \phi(f)k(a,t)d\nu(a)d\omega(t)\\ &= \int_M \phi(f)\phi\left(\int q_a d\nu(a)\right)d\omega(t)\\ &= [\, f, \int q_a d\nu_a \,] .\end{aligned}$$

Now remove the restriction that ν has compact support. Let K_n, $n = 1,2,\cdots$, be an exhaustion of R by compact subsets. That is, $R = \bigcup\{K_n\colon\ n = 1,2,\cdots\}$, and $K_n \subseteq \operatorname{Int} K_{n+1}$. Let

$$\nu_n = \nu|_{K_n} , \quad n = 1,2,\cdots .$$

Then

$$(14) \qquad \int q_a d\nu_n(a) \nearrow \int q_a d\nu(a) < \infty ,$$

pointwise. By the Harnack monotone convergence theorem, the right hand integral in 14 is harmonic, and thus lies in $h^1(R)$. From equation 3.5.5, we see that the integrals on the left in 14 converge to the right hand integral in $h^1(R)$ norm, and thus

$$[\, f, \int q_a dv_n(a)] \longrightarrow [\, f, \int q_a dv(a)\,] \,,$$

for each $f \in h^\infty(R)$. For all such f ,

$$\int f dv_n \longrightarrow \int f dv \,,$$

which establishes the theorem.

From the proof of theorem 2, we can also obtain

$$\phi\left(\int q_a dv(a)\right) = \int k_a dv(a) \quad \text{a.e.} \quad , \tag{15}$$

for all complex regular Baire measures v of finite total variation. We shall continue with the notation of the proof of theorem 2. Let

$$\chi_n = \chi_{K_n} \,, \quad n = 1,2,\cdots \quad ,$$

where χ_A temporarily denotes the characteristic function of A . By monotone convergence, for $z \in R$,

$$\int_M \int_R \chi_n(a) k(a,t) dv(a) d\omega_z(t) \nearrow \int_M \int_R k(a,t) dv(a) d\omega_z(t) \,. \tag{16}$$

The left hand side of 16 coincides with the left hand side of 14 thanks to equation 13. Thus the right hand side of 16 coincides with the right hand side of 14. By Fubini's theorem, the right hand side of 15 is finite a.e. and lies in $L^1(d\omega)$, which establishes equation 15.

4.5. In this section, and in this section only, $<\ ,\ >$ shall not represent the bilinear map introduced in definition 3.5.2. Rather, if E and F are vector spaces over $\mathbb{K}$, where $\mathbb{K}$ is either $\mathbb{R}$ or $\mathbb{C}$, then

$$< \ , \ >: \ E \times F \longrightarrow \mathbb{K}$$

shall be a bilinear form. We shall denote the dual system $(E,F,< \ , \ >)$ by $< E,F >$. We shall say $< E,F >$ is non-degenerate if $< x,y > = 0$ for all $y \in F$ implies $x = 0$, and if $< x,y > = 0$ for all $x \in E$ implies $y = 0$. Otherwise, we shall say the dual system $< E,F >$ is degenerate. We shall denote the weak topology on E relative to the dual system $< E,F >$ by $\sigma< E,F >$, and we shall let $\sigma < F,E >$ denote the weak topology on F. This and our other standard terminology concerning locally convex spaces are consistant with the terminology in [41].

By contrast with our temporary use of the symbol $< \ , \ >$, the symbol $[\ , \ ]$ shall continue to denote the bilinear form defined at the beginning of section 4.4. For $1 \leq p \leq \infty$, the dual system $[h^p(R),h^q(R)]$ is non-degenerate, as may be seen either from theorem 4.4.1, or by considering the functions $\{q_a: \ a \in R\}$.

Theorem 4.4.2 may be used to characterize the β dual of $h^\infty(R)$. The connection is provided by the following proposition. We shall omit the proof because it is identical, almost word for word, with the proof of the analogous proposition for $(H^\infty(R),\beta)$ given by Rubel and Shields in the case where R is a plane region [35].

Theorem 1: The locally convex space $(h^\infty(R),\beta)$ is complete. Further

A. The β dual of $h^\infty(R)$ is $M(R)$, the space of regular complex Baire measures of finite total variation on R, via the degenerate bilinear form

$$< f, \mu > = \int f d\mu ,$$

for $f \in h^\infty(R)$ and $\mu \in M(R)$.

B. The β bounded subsets of $h^\infty(R)$ are precisely the sets which are $h^\infty(R)$ norm bounded.

C. The topology of pointwise convergence, the topology of compact convergence, $\sigma\langle h^{\infty}(R), M(R)\rangle$, and β all agree on norm bounded subsets of $h^{\infty}(R)$.

D. The topologies β and $\sigma\langle h^{\infty}(R), M(R)\rangle$ are metrizable when restricted to norm bounded subsets of $h^{\infty}(R)$.

E. The β and $\sigma\langle h^{\infty}(R), M(R)\rangle$ convergent sequences in $h^{\infty}(R)$ are precisely the norm bounded pointwise convergent sequences.

F. The β and $\sigma\langle h^{\infty}(R), M(R)\rangle$ topologies have the same compact sets in $h^{\infty}(R)$, namely the pointwise closed norm bounded sets.

From the general theory of locally convex spaces, $\sigma\langle h^{\infty}(R), M(R)\rangle$ and β have the same bounded sets and the same closed convex sets in $h^{\infty}(R)$. Rubel and Ryff [37] and R. M. Dudley have shown that if $H^{\infty}(R)$ is infinite dimensional, then $(H^{\infty}(R), \beta)$ is not metrizable. In the opposite direction, we have the following generalization of a result of Rubel and Shields [35]:

<u>Theorem</u> 2: A convex subset of $h^{\infty}(R)$ is β closed if and only if it is sequentially closed.

Rubel and Shields established this result for $(H^{\infty}(R), \beta)$ where R was a plane region. Our proof depends on the Krein-Šmulian theorem, which we now quote (c.f. [41, p. 152]):

<u>Theorem</u> 3: A metrizable locally convex space E is complete if and only if a convex set $M \subseteq E'$ is $\sigma\langle E', E\rangle$ closed whenever $M \cap U^{o}$ is $\sigma\langle E', E\rangle$ closed for every 0 neighborhood U in E .

Here E' is the dual of E and $\langle f,g\rangle = g(f)$, for all $(f,g) \in E \times E'$. Further, $U^{o} = \{y \in E' : |\langle x,y\rangle| \leq 1 \text{ for all } x \in U\}$, is the polar of U .

<u>Proof of theorem</u> 2: Endow $M(R)$ with the total variation norm. Of course, $M(R)$ is now a Banach space. We may map $h^{\infty}(R)$ into $M(R)' = M'(R)$ via the bilinear form

$$\langle \mu, f\rangle = \int f\,d\mu ,$$

for all $(\mu,f) \in M(R) \times h^{\infty}(R)$. This mapping is 1 - 1, and the dual norm on $h^{\infty}(R)$ is simply the supremum norm, as may be seen by considering point evaluations. Thanks to theorem 1E, it suffices to show that a convex subset of $h^{\infty}(R)$ is $\sigma\langle h^{\infty}(R),M(R)\rangle$ closed if and only if it is $\sigma\langle h^{\infty}(R),M(R)\rangle$ sequentially closed. In this direction, let K be a convex subset of $h^{\infty}(R)$, and let B be the dual unit ball in $M'(R)$. Then $K \cap \lambda B$ is a convex bounded subset of $h^{\infty}(R)$ for each $\lambda \geq 0$. By theorem 1D, $K \cap \lambda B$ is $\sigma\langle h^{\infty}(R),M(R)\rangle$ closed if and only if $K \cap \lambda B$ is $\sigma\langle h^{\infty}(R),M(R)\rangle$ sequentially closed. Further, $\sigma\langle h^{\infty}(R),M(R)\rangle$ is simply the restriction of $\sigma\langle M'(R),M(R)\rangle$ to $h^{\infty}(R)$, so it suffices to show K is $\sigma\langle M'(R),M(R)\rangle$ closed in $h^{\infty}(R)$ if and only if $K \cap \lambda B$ is $\sigma\langle M'(R),M(R)\rangle$ closed in $h^{\infty}(R)$ for each $\lambda \geq 0$. But this, with the deletion of the phrase "in $h^{\infty}(R)$" , is simply the conclusion of the Krein-Šmulian theorem, since $M(R)$ is a Banach space. Thus it suffices to show $h^{\infty}(R)$ is weakly closed in $M'(R)$. However, for each $\lambda \geq 0$, $\lambda B \cap h^{\infty}(R)$ is a $\sigma\langle h^{\infty}(R),M(R)\rangle$ compact subset of $h^{\infty}(R)$ by theorem 1F, and thus is $\sigma\langle M'(R),M(R)\rangle$ compact, and hence is $\sigma\langle M'(R),M(R)\rangle$ closed in $M'(R)$. By the Krein-Šmulian theorem, $h^{\infty}(R)$ is $\sigma\langle M'(R),M(R)\rangle$ closed in $M'(R)$.

Let $(\Delta(0,1),\psi)$ be the universal convering surface of R . Using theorem 2, we see that a convex subset $K \subseteq h^{\infty}(R)$ is β closed if and only if the set $\{f \circ \psi\colon\ f \in K\}$ is β closed in $h^{\infty}(\Delta(0,1))$. Thus any theorem about β closed subspaces of $h^{\infty}(R)$ immediately gives us a theorem about β closed subspaces of the T automorphic functions in $h^{\infty}(\Delta(0,1))$, where T is the group of covering transformations. The main use of theorem 2, however, will be in the proof of the following theorem which more conveniently characterizes the β dual of $h^{\infty}(R)$:

<u>Theorem</u> 4: The dual of $(h^{\infty}(R),\beta)$ is $h^{1}(R)$ via the bilinear form [,] . Further, the dual system [,] is non-degenerate.

<u>Proof</u>: Let λ be a β continuous linear functional on $h^{\infty}(R)$. By theorem 1A, there exists a measure $\mu \in M(R)$ such that

$$\lambda(f) = \int f d\mu , \quad f \in h^\infty(R) .$$

By theorem 4.4.2,

$$\lambda(f) = [\ f, \int q_a d\mu(a)\ ] , \quad f \in h^\infty(R) .$$

Conversely, let $g \in h^1(R)$. Let $N = \{f \in h^\infty(R): [\ f,g\ ] = 0\}$. Let $f_n \in N$, $n = 1,2,\cdots$, and let $f_n \to f$ β as $n \to \infty$. The sequence f_n , $n = 1,2,\cdots$, is bounded, so it is $\sigma < h^\infty(R),h^1(R) >$ precompact by the Alaoglu theorem and $h^1(R) - h^\infty(R)$ duality. Let $f_{\sigma(n)}$, $n = 1,2,\cdots$ be any subsequence of $n = 1,2,\cdots$, and let f_α , $\alpha \in A$, be a subnet of $f_{\sigma(n)}$, $n = 1,2,\cdots$, converging $\sigma < h^\infty(R),h^1(R) >$ to some function $h \in h^\infty(R)$. The net f_α , $\alpha \in A$, also converges pointwise to f . Since $\sigma < h^\infty(R),h^1(R) >$ is finer than the topology of pointwise convergence, $h = f$. We have just proved that each subsequence of f_n , $n = 1,2,\cdots$, contains a subnet converging to f $\sigma < h^\infty(R),h^1(R) >$. Therefore, $f_n \to f$ $\sigma < h^\infty(R),h^1(R) >$. Consequently, $[\ f,g\ ] = 0$, i.e. $f \in N$. Thus N is β sequentially closed, and hence is β closed. We conclude that the linear functional

$$f \longmapsto [\ f,g\ ] , \quad f \in h^\infty(R) ,$$

is β continuous.

In the rest of this section, and throughout the remainder of this paper $<\ ,\ >$ shall once again denote the map introduced in definition 3.5.2. For each $h \in h^p(R)$, define the map

$$m_h : h^\infty(R) \longrightarrow h^p(R) ,$$

by $m_h(f) = < f,h >$, for all $f \in h^\infty(R)$. Suppose f_α , $\alpha \in A$, is a net in $h^\infty(R)$ converging to f β . By theorem 4, for each $g \in h^q(R)$, we have

$[< f_\alpha, h >, g] = [f_\alpha, < h, g >] \longrightarrow [f, < h, g >] = [< f, h >, g]$. We have just proved the first of the following two corollaries.

Corollary 1: For each $h \in h^p(R)$, the map m_h is $\beta - \sigma< h^p(R), h^q(R) >$ continuous.

Corollary 2: For each $h \in h^p(R)$, the map

$$\phi o m_h : \ h^\infty(R) \longrightarrow L^p(d\omega) ,$$

is $\beta - \sigma[L^p(d\omega), L^q(d\omega)]$ continuous.

In corollary 2, the bilinear form [,] is the standard bilinear form on $L^p \times L^q$. Corollary 2 follows immediately from corollary 1 and theorem 4.2.2, and is reminiscent of a lemma due to Havinson (c.f. [17, p. 77]). The proof of the first corollary may easily be generalized to show

Corollary 3: The space $(h^p(R), \sigma[h^p(R), h^q(R)])$ is an involutive $(h^\infty(R), \beta)$ locally convex module.

Chapter 5

Admissible Surfaces

5.1. It is imperative that we give some indication of the scope of the class of admissible surfaces, that is, the class to which our generalization of the Beurling-Rubel-Shields theorem applies. This we shall do in section 5.3, where we shall define the class of admissible surfaces and give examples. We shall postpone the proof that the examples given are actually admissible until chapter 8.

Throughout this paper, if $\mathbb{K}$ is a commutative ring and (X,A) is a space pair, then

$$H_n((X,A), \mathbb{K}) = H_n(X,A)$$

shall denote the nth singular homology group of (X,A) with coefficients in $\mathbb{K}$. Unless otherwise stated, we shall always employ non-argumented homology. The coefficient ring $\mathbb{K}$ will always be $\mathbb{Z}$, $\mathbb{R}$, or $\mathbb{C}$. In this chapter, $\mathbb{K} = \mathbb{Z}$. Fortunately, we shall never have to consider cohomology, so no confusion with Hardy class $H^p(R)$ can arise. Given a map between space pairs

$$f\colon (X,A) \longrightarrow (Y,B) ,$$

we shall often denote the map induced in homology by f rather than $f_{\#}$, again because no confusion with cohomology can arise.

Recall that f is a <u>global</u> <u>branch</u> of the l.m.m. F if f is meromorphic on R and $|f| = F$ on R. We wish to consider the question, when does F have a global branch? The question may be answered by considering the character of an l.m.m., a notion which, when phrased in the language of multiplicative multiple valued functions, goes back at least to Weyl's classic work, <u>Die Idee</u>

der Riemannschen Fläche [48].

To fix ideas, let $p \in R$ and let (p,f) be a function element belonging to F (c.f. section 2.4). We suppose F is not identically 0. Let γ be a closed arc in R with initial and terminal point p, and let (p,g) be the meromorphic continuation of (p,f) around γ. Then on U, the component of the intersection of the domains of f and g containing p, we have $|f| = |g|$. Thus f/g has only removable singularities on U, and by the maximum principle $f/g = c$, a constant of modulus 1. Since (p,f) may be continued along any arc, c depends only on the homotopy calss of γ [25, p. 113]. Let Γ be the homotopy class of γ and set $\theta(\Gamma) = c$. Clearly θ is a character on $\pi_1(R)$, the fundamental group of R. In other words, θ is a homomorphism of $\pi_1(R)$ into the circle group. In addition, θ does not depend on the choice of f. We shall call θ the character of F.

The character θ has a simple interpretation in terms of the universal covering surface of R, $(\Delta(0,1),\psi)$. Let T be the group of covering transformations and let $A\colon \pi_1(R) \longrightarrow T$ be the canonical isomorphism. Let f be a meromorphic function on R such that $F \circ \phi = |f|$. From the analytic continuation process used to construct θ, it follows that

$$\theta(\Gamma) = f \circ A(\Gamma)/f, \quad \Gamma \in \pi_1(R), \tag{1}$$

where we argue as before that $f \circ A(\Gamma)/f$ is constant by the maximum principle. Forelli recently considered characters from this point of view [9].

If F is of the form $\exp h$, where h is harmonic on R, then the character θ has a simple interpretation in terms of the periods of the differential $*dh$. Let γ be a piecewise smooth closed arc with initial and terminal point p, let Γ be the homotopy class of γ, and let U be a simply connected open set containing p. Choose any single valued harmonic conjugate h_* for h on U. The function element $(p, \exp(h+ih_*))$ belongs to F. From the analytic continuation process used to construct θ, we see that

$$\theta(\Gamma) = \exp i(h_{1*}-h_*),$$

where h_{1*} is the harmonic continuation of h_* around γ. It is standard that

$$h_{1*} - h_* = \int_\gamma *dh ,$$

(c.f. [31, pp. 31 and 73]). Thus

(2) $$\theta(\Gamma) = \exp(i\int_\gamma *dh).$$

Once again, let F be an arbitrary l.m.m.. Since the circle group is abelian, the kernel of θ contains the commutator subgroup of $\pi_1(R)$. However, $\pi_1(R)$ modulo its commutator subgroup is just $H_1(R)$. Thus we shall consider θ to be a character on $H_1(R)$. In particular, θ is independent of the choice of p.

Hereafter, if F is an l.m.m., we shall denote the character of F by θ_F. By convention, $\theta_0 = 1$. Clearly F has a global branch if and only if $\theta_F = 1$, the identity character. In addition, it is readily verified that θ is multiplicative on the multiplicative group of l.m.m.'s which are not $\equiv 0$. That is, $\theta_{FG}(\Gamma) = \theta_F(\Gamma)\theta_G(\Gamma)$ for all such l.m.m.'s F and G and all $\Gamma \in H_1(R)$.

5.2. We shall need a method for constructing analytic functions from l.a.m.'s. We would like the method to be well behaved from the point of view of ideal theory. To be well behaved, our method must at least turn bounded outer l.a.m.'s into exterior functions.

Definition 1: Let $\mathcal{X}$ be the group of characters (with values in the circle group) on $H_1(R)$. Then the family

(1) $$\delta_\chi , \quad \chi \in \mathcal{X},$$

is a system of multipliers for R if

A. δ_χ is an outer l.a.m. and $0 < \delta_\chi \leq 1$ for each $\chi \in \mathcal{X}$.

B. $\delta_1 = 1$ and δ_χ has character χ for each $\chi \in \mathcal{X}$.

C. For each integer $p \geq 1$, if a sequence of the form

$$\delta_{\chi(1,n)}\delta_{\chi(2,n)} \cdots \delta_{\chi(p,n)} , \quad n = 1,2,\cdots ,$$

converges pointwise to an l.a.m. with a global branch f, then f is exterior. Here $\chi(j,n) \in \mathcal{X}$ for $j = 1,2,\cdots,p$ and $n = 1,2,\cdots$.

Suppose now that R admits a system of multipliers. If F is an l.a.m. on R, and $\chi = \theta_F$, then $F\delta_{\chi^{-1}}$ is the modulus of an analytic function f on R.

Theorem 1: If F is bounded and outer, then f is exterior.

Proof: Set $\tau = \chi^{-1}$. Without loss of generality, assume $F = \exp(-H)$, where H is positive and quasi-bounded. Let H_n, $n = 1,2,\cdots$, be a sequence of positive bounded harmonic functions such that $H_n \nearrow H$. Then $F \exp H_n \nearrow 1$ pointwise on R. Set

$$\chi_n = \theta_{\exp H_n} , \quad n = 1,2,\cdots ,$$

and $\tau_n = \chi_n^{-1}$. Clearly $\delta_{\tau_n} \exp H_n$ has a global branch h_n for each n. Because the family $f h_n$, $n = 1,2,\cdots$, is bounded on R, we may assume, by passing to a subsequence if necessary, that the sequence $f h_n$, $n = 1,2,\cdots$, is β convergent to an analytic function g on R. However, $\lim_{n\to\infty} \delta_\tau \delta_{\tau_n} = \lim_{n\to\infty} F \exp H_n \delta_\tau \delta_{\tau_n} = |g|$, so g is β exterior. Then $\mathrm{Cl} f H^\infty(R) \supseteq \mathrm{Cl} g H^\infty(R) = H^\infty(R)$, so f is β exterior.

A condition weaker than C would guarantee the validity of theorem 1. The utility of condition C will become apparent in chapter 8 when we construct Riemann surfaces admitting systems of multipliers from other such surfaces.

Throughout this paper, we shall let

$$\delta U = dU + i{*}dU , \tag{2}$$

for each function U harmonic on R except for isolated singularities. Of course, δU is then analytic except possibly at the singularities of U. We are particularly interested in the meromorphic differential $\delta G(a,z)$ because we may express harmonic measure on M, the maximal ideal space of $h^{\infty}(R)$, with the help of $\delta G(a,z)$. The reason for this is that

$$(3) \qquad \frac{1}{-2\pi} *dG_S(a,z) = \frac{1}{-2\pi i} \delta G_S(a,z)$$

on ∂S for each $S \in J$. Here the notation is as in section 3.1 and the differentials are taken with respect to z. Note that the zeros of $\delta G(a,z)$ are simply the critical points of $G(a,z)$.

Definition 2: For $a \in R$, let Z_a be the set of zeros of $\delta G(a,z)$ and let $c(z)$ be the multiplicity of the zero at z for each $z \in Z_a$. Then R is G-bounded if a may be chosen so that

$$(4) \qquad \sum(c(z)G(a,z):\ z \in Z_a) < +\infty \ .$$

If R is G-bounded, we shall always assume the base point a_0 has been chosen to satisfy 4. By the Harnack monotone convergence theorem, if R is G-bounded, then

$$(5) \qquad \sum(c(z)G(a,z):\ z \in Z_{a_0})$$

is harmonic in a and finite for all $a \in R \sim Z_{a_0}$.

The G-boundedness condition for plane regions was first considered by us [26], and by Widom [50]. The termonology is ours.

5.3. We may now define the class of Riemann surfaces on which our generalization of the Beurling-Rubel-Shields theorem holds.

Definition 1: A regular Riemann surface R is admissible if it is G-bounded and admits a system of multipliers.

The reader should recall that R is regular if $\lim_{z\to\infty} G(a,z) = 0$ for each $a \in R$. Here $R \cup \{\infty\}$ is the Alexandroff compactification of R .

It is imperative that we give examples of admissible surfaces.

Theorem 1: Every non-compact finite Riemann surface is admissible.

The reader should recall that a non-compact finite Riemann surface R is simply the interior of a compact bordered Riemann surface. From this, it follows immediately that R is G-bounded and regular. Forelli [9] has recently given an elegant proof that R admits a system of multipliers by passing to the universal covering surface of R and using conditional expectations. In fact, he showed that A and C of definition 5.2.1 may be replaced by the stronger condition that for some $\varepsilon > 0$, $\varepsilon < \delta_\chi \leq 1$ for all $\chi \in \mathfrak{X}$. A classical proof of this, using a lemma in the proof of the Jacobi inversion theorem, is also possible (c.f. [48, p. 130] for a statement of this lemma).

An interesting class of examples, including surfaces of infinite connectivity, may be formed by punching holes, possibly infinite in number, from a larger G-bounded surface admitting a system of multipliers.

Definition 2: Let R_0 be a regular Riemann surface and let R be a region of R_0. Then (R_0,R) is a Blaschke pair if each component of $R_0 \sim R$ is a compact continuum, and

A. For each component A of $R_0 \sim R$, there exists a point $a_A \in A$ such that

(1) $$\sum(G_0(a_0,a_A) : A \text{ is a component of } R_0 \sim R) < \infty .$$

(Here, and throughout the rest of this section, G_0 shall be the Green's function for R_0 .)

B. The components of $R_0 \sim R$ are open in $R_0 \sim R$.

C. For each component A of $R_0 \sim R$, $V \sim A$ is connected for each connected neighborhood V (in R_0) of A.

Condition C is not vacuous. Since A is open in $R_0 \sim R$ and R_0 is locally connected, there exist open connected neighborhoods of A separating A from $R_0 \sim (R \cup A)$.

In our thesis, we considered Blaschke pairs (R_0,R), where R_0 was a Jordan region in the plane [26]. We called such a region R a Blaschke region. The reason is that the proto-type of a Blaschke region is the unit disk with (possibly infinitely many) disjoint circular holes, centered at the zeros of a convergent Blaschke product, punched out. Blaschke regions were first considered by Voichick [46].

Let R_0 be a regular Riemann surface and let (R_0,R) be a Blaschke pair. Clearly R is regular.

Theorem 2: If R_0 admits a system of multipliers, then so does R.

If R admits a system of multipliers, we may apply some remarkable recent results of Widom's to conclude that R is G bounded [50], [51]. However, the following result also suffices for our purposes, and its proof is of independent interest.

Theorem 3: If R_0 is G-bounded, then so is R.

We shall omit the proofs of theorems 2 and 3 until chapter 8. However, these theorems do indicate the scope of the class of admissible surfaces. For example, if we start with a non-compact finite Riemann surface R_0 and construct a chain of Blaschke pairs

$$(R_0, R_1)\ ,\ (R_1, R_2)\ ,\ \cdots,\ (R_{n-1}, R_n),$$

then the terminal surface of the chain, R_n, is admissible. The surface R_n may be of infinite connectivity, in fact R_n is constructed from R_0 by deleting finitely or infinitely many disjoint compact continua. Extremely elementary examples show that even if $R_0 = \Delta(0,1)$ and $n = 2$, R_n need not be a Blaschke region.

Unfortunately, we cannot construct admissible surfaces with infinitely many handles by this process.
The following conjecture, if true, would produce such surfaces:

Conjecture 1: Let R be an admissible surface. Then each normal covering surface of R is also admissible.

Chapter 6

Cauchy-Read Theorems

6.1. In this chapter, we shall present two of the three major results in this paper. They are theorem 6.2.1, which is a version of the Cauchy integral formula involving $\phi(f)$ for suitable f, and a converse to this theorem, theorem 6.3.1, which is a generalization of the theorem of A. H. Read [32]. Throughout this chapter and the next, unless otherwise stated, R shall be an admissible surface admitting the system of multipliers 5.2.1. Also, we shall return to the conventions of chapters 3 and 4. In particular, $1 \leq p \leq \infty$, $1 \leq q \leq \infty$, and $1/p + 1/q = 1$.

Let f be a meromorphic function on R of bounded characteristic. Suppose $f \not\equiv 0$ or ∞. From the comments at the end of section 2.3, we have

$$|f| = F_1/F_2 , \tag{1}$$

where F_1 and F_2 are bounded l.a.m.'s. Applying the remarks following definition 5.2.1, we see that for appropriate l.a.m.'s δ_1 and δ_2 chosen from the system of multipliers, we have

$$F_1\delta_1 = |f_1| , \quad F_2\delta_2 = |f_2| , \tag{2}$$

where f_1 and $f_2 \in H^\infty(R)$. Since

$$1 = \theta_{|f|} = \theta_{F_1}(\theta_{F_2})^{-1} , \tag{3}$$

we know $\delta_1 = \delta_2$. Thanks to the maximum principle,

(4) $$f = cf_1/f_2 ,$$

for a suitably chosen constant c . We shall always suppose f_1 has been chosen so $c = 1$. We have just given the proof of

Lemma 1: Each meromorphic function of bounded characteristic on R is the quotient of bounded analytic functions.

Thanks to lemma 1 and theorem 4.3.3, we have

Definition 1: Let f be a meromorphic function of bounded characteristic on R . Then

(5) $$\phi(f) = \phi(f_1)/\phi(f_2) ,$$

for $f \not\equiv \infty$, where f_1 and f_2 are any pair of functions lying in $H^\infty(R)$ for which equation 4 holds. Also $\phi(\infty) \equiv \infty$.

Using theorems 4.2.2 and 4.3.3, it is easy to show that $\phi(f)$ is independent of the decomposition 4, and that $\phi(f)$ is defined and finite a.e., unless $f \equiv \infty$. In addition, it is easy to show that the extension of ϕ given by definition 1 is consistent with the previous extension to $H^p(R)$. Finally, we have

Lemma 2: Let f be a meromorphic function of bounded characteristic on R . If $\phi(f) = 0$ on a set of positive measure, then $f \equiv 0$. If $\phi(f) = \infty$ on a set of positive measure, then $f \equiv \infty$.

Although the proof is similar to the others we have just omitted, we shall give it for reference.

Proof: Suppose $\phi(f) = 0$ on a set of positive measure. Consider the decomposition 4. We have $\phi(f_1) = 0$ on a set of positive measure. Since $f_1 \in H^\infty(R)$, we may apply theorem 4.3.3 to conclude that $f_1 \equiv 0$. The function $f_2 \not\equiv 0$, since $f \not\equiv \infty$. Therefore, $f \equiv 0$. The case where $\phi(f) = \infty$ on a set of positive measure is subsumed under our previous remarks.

6.2. Let S be the $SP(R)$ function 5.2.5 and let

$$g = \exp(-S) . \tag{1}$$

Clearly g is a generalized Blaschke product. Let δ be the l.a.m. chosen from the system of multipliers on R such that $g\delta$ has character 1. Let

$$B \in H^\infty(R) \tag{2}$$

be a global branch of $g\delta$. Note that $|B| \le 1$, the inner factor of $|B|$ is g, and both g and B have exactly the same zeros, counting multiplicity, as the meromorphic differential $\delta G(a_0,z)$.

<u>Lemma</u> 1: Let f be a meromorphic function on R. Suppose $g|f|$ admits a harmonic majorant. Then f is of bounded characteristic on R and $\phi(f) \in L^1(d\omega)$. If $g|f|$ is bounded, then $\phi(f) \in L^\infty(d\omega)$.

This lemma allows us to state our version of the Cauchy integral formula. To motivate our version, let $S \in J$ and consider the meromorphic differential on $\mathrm{Cl}S$,

$$\frac{1}{-2\pi i}\, \delta G_S(a,z) . \tag{3}$$

From equation 5.2.3, we see that the differential 3 on ∂S is simply harmonic measure on ∂S at a. Let f be analytic on S except for poles at the zeros of $\delta G_S(a_0,z)$, and continuous in the chordal metric on $\mathrm{Cl}S$. Suppose the orders of the poles of f have been chosen so that the differential $f\delta G_S(a_0,z)$ is analytic on S except for the pole at a_0. Then by the calculus of residues,

$$\frac{1}{-2\pi i}\int_{\partial S} f\delta G_S(a_0,z) = f(a_0) . \tag{4}$$

This is clearly an integral formula of Cauchy type. Our generalization is

<u>Theorem</u> 1: Let f be a meromorphic function on R. Suppose $g|f|$ has a harmonic majorant. Then

$$\int_M \phi(f)d\omega = f(a_0) . \tag{5}$$

As will be clear from its proof, theorem 1 may be generalized to give a calculus of residues theorem. It will also be clear from sections 3.3 and 3.6 that theorem 1 may be phrased in terms of the Martin boundary. That is, if f satisfies the hypotheses of theorem 1, then

$$\int_{\Delta_1} \hat{f}d\omega_{a_0} = f(a_0) ,$$

where the notation of sections 3.3 and 3.6 has been used. Despite the sketchy nature of these remarks, we shall not persue either of these matters further in this paper.

<u>Proof of lemma</u> 1. Suppose $g|f|$ admits a harmonic majorant. Then $g|f|$, and hence f, are of bounded characteristic. Write

$$\ln g|f| = h + s ,$$

where h is quasi-bounded, s is inner, and $s \leq 0$. This is possible by theorem 2.5.3. We have

$$|f| = \frac{\delta \exp(h+s)}{\delta g} = \frac{\delta \exp(h+s)}{|B|} .$$

Thus $\delta \exp(h+s)$ has a global branch $F \in H^1(R)$. The outer factor of $|F|$ is $\delta \exp h$. Thanks to equation 4.3.5,

$$|\phi(F)| = \exp(\phi(h)+\phi(\ln \delta)) \quad \text{a.e.} \quad .$$

Similarly,

$$|\phi(B)| = \exp \phi(\ln \delta) ,$$

so

$$|\phi(f)| = \left|\frac{\phi(F)}{\phi(B)}\right| = \exp \phi(h) \quad \text{a.e.} \quad .$$

As in the derivation of equation 4.3.5, we obtain

$$\exp \phi(h) = \phi(\text{L. H. M. } g|f|) \in L^1(d\omega) .$$

Clearly, if $g|f|$ is bounded, so is $\phi(f)$ a.e..

For $S \in J$, it will be convenient to replace the differential $\delta G_S(a_0,z)$ with the differential $\delta G(a_0,z)$ when discussing harmonic measure on ∂S. It is possible to do this for sufficiently many $S \in J$ due to the following lemma. Here, and in the next few paragraphs, R shall simply be a regular Riemann surface.

<u>Lemma</u> 2: There exists an exhaustion of R by Jordan subregions of the form

(6) $$\{z\colon\ G(a_0,z) > \alpha\} .$$

Call $\alpha > 0$ a <u>critical value</u> of $G(a_0,z)$ if $G(a_0,z)$ has a critical point in the set $\{z: G(a_0,z) = \alpha\}$. Whether or not $\alpha > 0$ is a critical value, it is utterly standard that the set

(7) $$R^\alpha = \{z: G(a_0,z) \geq \alpha\} .$$

is connected. (C.f. [14]. A proof may also be based on lemmas 8.1.1 and 8.1.2). The boundary of R^α is

(8) $$\partial R^\alpha = \{z: G(a_0,z) = \alpha\},$$

which is the disjoint union of analytic curves if α is not a critical value of $G(a_0,z)$. In fact, we may cover ∂R^α by parametric disks (V,ψ) such that

(9) $$V \cap R^\alpha = \psi^{-1}(\Delta(0,1) \cap \{\operatorname{Im} z \geq 0\})$$

$$V \cap (R \sim R^\alpha) = \psi^{-1}(\Delta(0,1) \cap \{\operatorname{Im} z < 0\})$$

$$V \cap \partial R^\alpha = \psi^{-1}(-1,1) ,$$

provided $\alpha > 0$ is not a critical value of $G(a_0,z)$. By the regularity of R , we know R^α is compact. Thus R^α is a closed Jordan subregion of R if $\alpha > 0$ is not a critical value of $G(a_0,z)$. We may now give the proof of lemma 2.

Proof: It suffices to show that there are enough non critical values of $G(a_0,z)$. But by the identity theorem for meromorphic differentials, $\delta G(a_0,z)$ has only finitely many zeros in each of the compact R^α . Thus the critical values of $G(a_0,z)$ form a discrete subset of $(0,\infty)$.

From now on, we shall redefine J to be the directed set of Jordan subregions of the form 6. For each $S \in J$,

(10) $$G_S(a_0,z) = G(a_0,z) - \alpha ,$$

for a suitable constant α . Thus

$$\delta G_S(a_0,z) = \delta G(a_0,z) \tag{11}$$

on ∂S , for each $S \in J$.

We now reimpose the assumption that R is admissible. Throughout this section, we shall let z_n , $n = 1,2,\cdots$ be a univalent enumeration of the zeros of $\delta G(a_0,z)$. We shall let c_n be the order of the zero at z_n for each n . Temporarily fix n and let

$$F(z) = \exp(- \textstyle\sum_j^n c_j G(z_j,z)) \ , \quad z \in R \ .$$

Choose ε from the system of multipliers so that $F\varepsilon$ has a global branch $D \in H^\infty(R)$. Choose τ from the system of multipliers so that $\varepsilon\tau$ has a global branch $u \in H^\infty(R)$.

Lemma 3: Let f be a meromorphic function on R such that $F|f|$ has a harmonic majorant. Then equation 5 holds.

Note that $g \leq F$, so $g|f|$ has a harmonic majorant. By lemma 1, $\phi(f)$ is integrable, and thus there exists a unique $h \in h^1(R)$ such that $\phi(f) = \phi(h)$ a.e.. Also $Df \in H^1(R)$ and

$$\phi(Df) = \phi(D)\phi(f) = \phi(D)\phi(h) \quad \text{a.e.} \ .$$

Thanks to theorem 4.2.2, $Df = < D,h >$. We are now ready to begin the proof of lemma 3.

Proof: Let $s \in H^\infty(R)$. We have

(12)
$$\lim_{S\in J} \frac{1}{-2\pi i}\int_{\partial S} |suf - suh|\,\delta G(a_0,z)$$

$$= \lim_{S\in J} \frac{1}{-2\pi i}\int_{\partial S} \frac{|s||u|}{|D|}\,|Df - Dh|\,\delta G(a_0,z)$$

$$= \lim_{S\in J} \frac{1}{-2\pi i}\int_{\partial S} \frac{\tau|s|}{F}\,|< D,h > - Dh|\,\delta G(a_0,z)$$

$$= 0 \quad .$$

Here we have used theorem 3.5.1. In addition, we have used the regularity of R to conclude that $\lim_{z\to\infty} F(z) \neq 0$.

For each $S \in J$, we may apply the calculus of residues to obtain

(13)
$$\frac{1}{-2\pi i}\int_{\partial S} suf\delta G(a_0,z) = suf\Big|_{a_0} \quad .$$

From equations 12 and 13, we conclude that

$$suf\Big|_{a_0} = \lim_{S\in J} \frac{1}{-2\pi i}\int_{\partial S} suh\delta G(a_0,z) = [\ su,h\ ] = \int \phi(su)\phi(f)d\omega \ ,$$

for all $s \in H^\infty(R)$. Thanks to definition 5.2.1C, u is exterior. Thus there exists a net s_α , $\alpha \in A$, in $H^\infty(R)$ such that $s_\alpha u \to 1$ β . Applying corollary 4.5.4.2, we obtain

$$f(a_0) = \lim_{\alpha\in A} s_\alpha uf\Big|_{a_0} = \lim_{\alpha\in A} \int \phi(s_\alpha u)\phi(f)d\omega = \int \phi(f)d\omega \quad .$$

We now may give the proof of theorem 1.

<u>Proof</u>: For $n = 1,2,\cdots$, let

$$g_n(z) = \sum_{j=n+1}^{\infty} c_j G(z_j, z) \quad , \quad z \in R .$$

Choose δ_n from the system of multipliers so that $g_n\delta_n$ has a global branch B_n. Since $|B_n| \le 1$ for each n, an application of theorem 4.5.1 shows that a subsequence $B_{\sigma(n)}$, $n = 1,2,\cdots$ converges β to $C \in H^\infty(R)$. Due to the convergence of the series 5.2.4, $g_{\sigma(n)} \to 1$ β as $n \to \infty$, so $\delta_{\sigma(n)} \to |C|$ pointwise. Thanks to definition 5.2.1, C is exterior. Clearly, for each n, $B_n f$ satisfies the hypotheses of lemma 3. Thus

$$Cf\Big|_{a_0} = \lim_{n\to\infty} B_{\sigma(n)} f\Big|_{a_0} = \lim_{n\to\infty} \int \phi(B_{\sigma(n)})\phi(f)d\omega = \int \phi(C)\phi(f)d\omega .$$

As in the proof of lemma 3, we may conclude that equation 5 holds by appealing to the exteriority of C.

6.3. In the course of a study of conformal mapping problems, A. H. Read established the following theorem [32]: Let $R \cup \Gamma$ be a compact bordered Riemann surface with interior R and border Γ. Let $f \in L^p(\frac{1}{-2\pi i}\, \delta G(a_0,z))$, and suppose

$$\int_\Gamma f\nu = 0 ,$$

for each differential 1-form ν analytic on $R \cup \Gamma$. Then there exists a function $h \in H^p(R)$ such that $h^* = f$ a.e. on Γ. Here, h^* is the non-tangential boundary limit function of h. We have the following generalization of the theorem of A. H. Read. Here R is an arbitrary admissible surface.

Theorem 1: Let $f \in L^p(d\omega)$. Suppose

$$\int \phi(h) f d\omega = 0 ,$$

for each meromorphic function h on R such that $g|h|$ is bounded and $h(a_0) = 0$. Then there exists a function $F \in H^p(R)$ such that $\phi(F) = f$ a.e. on M.

Theorem 1, as well as Read's original theorem, are converses to the Cauchy integral theorem, much as Morera's theorem is. We are indebted to J. A. Jenkins for the observation that Read's theorem for $\mathrm{Cl}\Delta(0,1)$ is actually a consequence of Morera's theorem. Our proof of theorem 1 is an extension of his observation.

As will be clear from its proof, theorem 1 may be phrased in terms of the Martin boundary by using sections 3.3 and 3.6. That is, if $f \in L^p(d\omega_{a_0})$ and if

$$\int_{\Delta_1} \hat{h} f d\omega_{a_0} = 0$$

for each meromorphic function h satisfying the hypotheses of theorem 1, then there exists $F \in H^p(R)$ such that $\hat{F} = f$ a.e. on Δ_1. Here, we have used the notation of sections 3.3 and 3.6. Despite the sketchy nature of these observations, we shall not pursue them further in this paper.

In order to carry out the proof of theorem 1, we shall need several results which will be of use later. We shall, therefore, postpone the proof of our generalization of Read's theorem until the beginning of section 6.4. Throughout the remainder of this section, R shall simply be a regular Riemann surface.

Recall that $G(a,z)$ is harmonic in each variable, symmetric, and jointly continuous on $R \times R \sim \Delta_R$. Throughout this section, $\Delta_R = \{(z,z): z \in R\}$. It is also standard that

$$(1) \qquad \int G(a,z)d\nu(a)$$

is harmonic in z on $R \sim K$ for each regular Baire measure ν with compact support K (c.f. [5, pp. 32-33]).

Let A be a compact subset of R, and let (V,ψ) be a parametric disk in R such that $\mathrm{Cl}V \cap A = \phi$. Then the difference quotients

$$\frac{1}{h}(G(a,\psi^{-1}(x+h+iy)) - G(a,\psi^{-1}(x+iy))) ,$$

(2)

$$\frac{1}{h}(G(a,\psi^{-1}(x+i(y+h))) - G(a,\psi^{-1}(x+iy))) ,$$

converge uniformly on $A \times \Delta(0,1)$ to

(3) $$\frac{\partial}{\partial x} G(a,\psi^{-1}(x+iy)) \text{ and } \frac{\partial}{\partial y} G(a,\psi^{-1}(x+iy)) ,$$

respectively. The argument establishing this is a standard one for pluriharmonic functions, but we shall give it for reference. From the definition of a parametric disk [1, pp. 131-132], we conclude there is a larger parametric disk (W,ψ) such that $\mathrm{Cl}V \subseteq W$ and $W \cap A = \phi$. Assume $\psi(V) = \Delta(0,1)$, $\psi(W) = \Delta(0,r_W)$, and let $U = \psi^{-1}(\Delta(0,r))$, where $1 < r < r_W$. The function $G(a,z)$ is harmonic in z and continuous on $A \times \Delta(0,r_W)$. Consider the Poisson kernel $K(z,re^{i\theta})$ for $\Delta(0,r)$, and the difference quotient

(4) $$D(x+iy,h,re^{i\theta}) = \frac{1}{h}(K(x+h+iy,re^{i\theta}) - K(x+iy,re^{i\theta})) .$$

Then the first of the difference quotients 2 is given by

(5) $$\frac{1}{2\pi}\int_0^{2\pi} G(a,\psi^{-1}(re^{i\theta}))D(x+iy,h,re^{i\theta})d\theta$$

on $A \times \Delta(0,1)$. However, the integral 5 converges uniformly to the first of the partial derivatives 3, thanks to the uniform boundedness of $G(a,z)$ on $A \times \Delta(0,r)$ and the uniform convergence of the difference quotients 4 to

$$\frac{\partial}{\partial x} K(x+iy,re^{i\theta}) ,$$

on $\Delta(0,1) \times \partial\Delta(0,r)$. Incidently, we have also shown,

$$(6) \qquad \frac{\partial}{\partial x} G(a,\psi^{-1}(x+iy)) = \frac{1}{2\pi}\int_0^{2\pi} G(a,\psi^{-1}(x+iy)\,\frac{\partial}{\partial x} K(x+iy,re^{i\theta})d\theta \quad ,$$

and a similar equation holds for the other partial derivative.

Now let (V,ψ) be an arbitrary parametric disk in R, and let $\Delta_\psi = \{(a,z): \ \psi(a) = z \ , \ a \in V\}$. From equation 6 and from the equation

$$(7) \qquad \frac{\delta G(a,\psi^{-1}(z))}{dz} = \left(\frac{\partial}{\partial x} - i\,\frac{\partial}{\partial y}\right)G(a,\psi^{-1}(x+iy)) \ ,$$

we immediately obtain

<u>Lemma</u> 1: The partial derivatives 3, as well as the function 7, are all jointly continuous on $R \times \Delta(0,1) \sim \Delta_\psi$.

<u>Lemma</u> 2: Let K be a compact subset of R . Let μ be a regular complex Baire measure with support contained in K . Let C be a piecewise smooth arc disjoint from K . Then

$$\int_C f{*}d\int_K G(a,z)d\mu(a) = \int_K\int_C f{*}dG(a,z)d\mu(a) \ ,$$

for each complex valued function f continuous on C .

<u>Proof</u>: Since the image of C is compact, it suffices to assume C is contained in a single parametric disk (V,ψ) whose closure is disjoint from K . By Fubini's theorem, we have

$$\int_{\psi o C} f\left(\frac{\partial}{\partial y}\int_K G(a,\psi^{-1}(x+iy))d\mu(a)\right)dx$$

$$= \int_{\psi o C}\int_K f\,\frac{\partial}{\partial y}\,G(a,\psi^{-1}(x+iy))d\mu(a)dx$$

$$= \int_K\int_{\psi o C} f\,\frac{\partial}{\partial y}\,G(a,\psi^{-1}(x+iy))dxd\mu(a) \ .$$

Here the interchange of integration and differentiation is justified by the uniform convergence of the difference quotients 2. The dy terms are treated similarly.

Let $S \in J$ and let T be the Schottky double of the compact bordered surface $S \cup \partial S$. Since $G_S(a,z) = 0$ on $S \times \partial S$, the function $z \to G_S(a,z)$ has a harmonic extension to T, which we shall also denote by $G_S(a,z)$. We have

$$G_S(a,\tau(z)) = -G_S(a,z) ,$$

where $\tau\colon T \longrightarrow T$ is the standard conformal map carrying $S \cup \partial S$ onto $T \sim S$. Thus $G_S(a,z)$ is jointly continuous on $S \times T \sim \Delta_T$, where $\Delta_T = \{(z,z)\colon z \in T\}$. From this, we conclude that lemma 1 holds if (V,ψ) is a parametric disk in T. Of course, T must be substituted for R and $G_S(a,z)$ for $G(a,z)$ in the statement of lemma 1. In particular, we have the first assertion of

Lemma 3: The function

$$\delta G_S(a,z)/\delta G(a_0,z)$$

is continuous on $S \times \partial S$ for each $S \in J$. Furthermore, if A is a compact subset of R and if (V,ψ) is a parametric disk in R such that $\mathrm{Cl}V \cap A = \phi$, then

$$(8) \qquad \lim_{S\in J} \frac{\delta G_S(a,\psi^{-1}(z))}{dz} = \frac{\delta G(a,\psi^{-1}(z))}{dz}$$

uniformly on $A \times \Delta(0,1)$.

Proof: We know that $G_S(a,z) \nearrow G(a,z)$, $S \in J$. By Dini's theorem, $G_S(a,z) \nearrow G(a,z)$, $S \in J$, uniformly on compact subsets of $R \times R \sim \Delta$. Thanks to equation 6, equation 8 holds.

6.4. We are now in a position to give the proof of theorem 6.3.1, our generalization of Read's theorem. The proof depends on two key lemmas, which we shall state presently. We reimpose the hypothesis that R is an admissible surface. Throughout this section, we shall let Γ be the set of all ordered pairs $(\gamma,(V,\psi))$ satisfying the following conditions: The pair (V,ψ) is a parametric disk such that $a_0 \notin \mathrm{Cl}V$, and γ is a closed oriented curve lying in V such that $\psi \circ \gamma$ is rectifiable. For notational convenience, we shall suppress (V,ψ) when referring to quantities depending on $(\gamma,(V,\psi))$. For example, we shall let $\underline{l}(\gamma)$ be the length of $\psi \circ \gamma$.

<u>Lemma</u> 1: For each $(\gamma,(V,\psi)) \in \Gamma$, the function

$$(1) \qquad \frac{1}{B(z)} \int_\gamma B(z) \frac{\delta G(a,z)}{\delta G(a_0,z)} \, d\psi(a) \; , \quad z \in R \sim \mathrm{Cl}V \; ,$$

possesses a meromorphic extension P_γ to all of R. Furthermore, $g|P_\gamma|$ is bounded and $P_\gamma(a_0) = 0$.

<u>Lemma</u> 2: For each $(\gamma,(V,\psi)) \in \Gamma$,

$$\phi(P_\gamma) = \int_\gamma k_a \, d\psi(a) \quad \text{a.e.} \; ,$$

where k_a is the continuous function agreeing with the function 4.4.1 a.e.

<u>Proof</u> <u>of</u> <u>Theorem</u> 6.3.1: Let $F = \phi^{-1}(f)$, where ϕ is either the map 4.2.2 or 4.2.5. Then $F \in h^p(R)$ and $\phi(F) = f$ a.e.. Thanks to theorem 4.4.2 and equation 4.4.13,

$$\int_\gamma F d\psi(a) = [\, F, \int_\gamma q_a d\psi(a) \,] = \int_M \phi(F)\Big|_t \left(\int_\gamma k_a(t) d\psi(a)\right) d\omega(t) \; ,$$

for each $(\gamma,(V,\psi)) \in \Gamma$. Appealing to lemmas 1 and 2, we obtain

(2) $$\int_\gamma F d\psi = \int_M f\phi(P_\gamma)d\omega = 0 ,$$

for each $(\gamma,(V,\psi)) \in \Gamma$. By Morera's theorem, F is analytic, except possibly at a_0 . Since F is harmonic on R , and hence bounded near a_0 , F is also analytic at a_0 .

We still must establish lemmas 1 and 2. Let $(\gamma,(V,\psi)) \in \Gamma$. By appealing to lemma 6.3.1, we see that the function

(3) $$B(z)\ \delta G(a,z)/\delta G(a_0,z) ,$$

is jointly continuous on $R \times R \sim \Delta$. The function 3 is also analytic in z , thanks to the remarks immediately following formula 6.2.2. A simple application of Fubini's theorem together with Morera's theorem shows that the function

(4) $$\int_\gamma B(z) \frac{\delta G(a,z)}{\delta G(a_0,z)} d\psi(a) , \quad z \in R \sim \text{Cl}V ,$$

is analytic on $R \sim \text{Cl}V$. Because $\delta G(a_0,z)$ has a pole at a_0 , the function 4 is 0 at $z = a_0$.

<u>Proof of Lemma</u> 1: First, we shall show that the function 4 may be extended analytically to all of R . Let (W,ψ) be a parametric disk such that $\text{Cl}V \subseteq W$. Let $\psi(V) = \Delta(0,1)$, $\psi(W) = \Delta(0,s)$, and suppose $1 < r < s$. For convenience, temporarily write $G(a,z)$ for $G(\psi^{-1}(a),\psi^{-1}(z))$, $G(a_0,z) = G(a_0,\psi^{-1}(z))$, $\Delta = \{(z,z)\colon\ z \in \Delta(0,s)\}$, and $\gamma = \psi \circ \gamma$. Recall that

$$G(a,z) = h(a,z) - \ln|z-a| ,$$

where $h(a,z)$ is harmonic on each variable and symmetric on $\Delta(0,s) \times \Delta(0,s)$, and jointly continuous on $\Delta(0,s) \times \Delta(0,s) \sim \Delta$. By writing $h(a,z)$ as a Poisson integral, we see that $h(a,z)$ is jointly continuous on $\Delta(0,r) \times \Delta(0,r)$. As in the proof of lemma 6.3.1, the function $\delta h(a,z)/dz$ is not only analytic in z, but also is jointly continuous on $\Delta(0,r) \times \Delta(0,r)$. Thus the function

$$(5) \qquad \int_\gamma \frac{\delta h(a,z)}{dz}\, da \ , \quad z \in \Delta(0,r) \ ,$$

is analytic on $\Delta(0,r)$. An elementary calculation shows

$$\frac{\delta \ln|z-a|}{dz} = \frac{1}{z-a} \ ,$$

on $\Delta(0,r) \times \Delta(0,r)$. However,

$$\int_\gamma \frac{da}{z-a} = 0 \ , \quad z \in \Delta(0,r) \sim \Delta(0,1) \ ,$$

since the winding number of γ with respect to each point in the unbounded component of $\mathbf{C} \sim \gamma$ is 0. Thus

$$(6) \qquad \int_\gamma \frac{G(a,z)}{dz}\, da = \int_\gamma \frac{h(a,z)}{dz}\, da \ ,$$

on $\Delta(0,r) \sim \Delta(0,1)$. From equation 6 and the analyticity of the function 5, we conclude that the function 4 possesses an analytic extension to all of R. Consequently, the function 1 possesses a meromorphic extension P_γ to all of R.

At this point, we shall interrupt the proof of lemma 1 briefly. Let $S \in J$. Thanks to formulas 6.2.3 and 6.2.11, for each $a \in S$, the function

$$\delta G_S(a,z)/\delta G(a_0,z) \ , \quad z \in \partial S \ ,$$

is the Radon-Nikodym derivative of harmonic measure at a on ∂S with respect to harmonic measure at a_0 on ∂S. As in the derivation of inequality 4.4.2, we conclude

(7) $$\frac{\delta G_S(a,z)}{\delta G(a_0,z)} \geq 0 ,$$

on $S \times \partial S$. Further, if K is a compact subset of S ,

(8) $$\lambda(K,S) \leq \frac{\delta G_S(a,z)}{\delta G(a_0,z)} \leq \mu(K,S) ,$$

on $K \times \partial S$. Here, and throughout the rest of this section, $\lambda(K,S)$ and $\mu(K,S)$ shall denote the lower and upper Harnack constants for K as a subset of S .

<u>Proof of Lemma 1 continued</u>: For each $S \in J$ such that $\mathrm{Cl}V \subseteq S$, define $P_{S,\gamma}$ analogously to P_γ , by replacing $G(a,z)$ by $G_S(a,z)$ and R by S in formula 1. Clearly $g|P_{S,\gamma}|$ is an l.a.m. on S . From lemma 6.3.3 and the equation

(9) $$P_{S,\gamma}(z) = \int_\gamma \frac{\delta G_S(a,z)}{\delta G(a_0,z)}\, d\psi(a) ,$$

for all $z \in S$ which are not critical points of $\delta G(a_0,z)$, we conclude that $g|P_{S,\gamma}|$ has a continuous extension to $\mathrm{Cl}S$. Equation 9 also allows us to extend $P_{S,\gamma}$ to $\mathrm{Cl}S$ so that the extension, also denoted by $P_{S,\gamma}$, is continuous in a neighborhood of ∂S . By inequality 8,

$$|P_{S,\gamma}| \leq \mu(S)\underline{1}(\gamma) ,$$

on ∂S , where $\mu(S) = \mu(\gamma, S)$. By the maximum principle,

(10) $$g|P_{S,\gamma}| \leq \mu(S)\underline{1}(\gamma)$$

on $\mathrm{Cl}S$.

Let $S_0 \in J$ be such that $\mathrm{Cl}V \subseteq S_0$. Let $S \in J$ be such that $\mathrm{Cl}S_0 \subseteq S$. An application of lemma 6.3.3 and the Heine-Borel theorem to $\gamma \times \partial S$ shows that

$$\lim_{T \in J} P_{T,\gamma} = P_\gamma \tag{11}$$

uniformly on ∂S . From inequality 10, we obtain

$$g|P_\gamma| \leq \mu(S_0)\underline{l}(\gamma) \quad ,$$

on ∂S , and thus on $\mathrm{Cl}S$. Here we have used the inequality

$$\mu(T) \leq \mu(S) \leq \mu(S_0) \quad , \tag{12}$$

for $S_0 \subseteq S \subseteq T$ and S , $T \in J$. Since S was an arbitrary region in J containing S_0 ,

$$g|P_\gamma| \leq \mu(S_0)\underline{l}(\gamma)$$

on R . Because the function 4 vanishes at a_0 , so does P_γ . This completes the proof of lemma 1.

It will be convenient to introduce the notation

$$< f,g >_S\Big|_a = \frac{1}{-2\pi i}\int_{\partial S} fg\delta G_S(a,z) \quad ,$$

for each $S \in J$, $a \in S$, and pair of functions f,g harmonic on S and continuous on $\mathrm{Cl}S$. Extend $< f,g >_S$ to $\mathrm{Cl}S$ by continuity. We shall also denote the extension by $< f,g >_S$. Suppose $f \in h^p(R)$ and $g \in h^q(R)$. By theorem 3.5.1,

$$\lim_{S \in J} \frac{1}{-2\pi i}\int_{\partial S} |< f,g >_S - < f,g >|\,\delta G_S(a,z) = 0 \quad ,$$

for each $a \in R$. From the Harnack inequalities, we conclude that

(13) $$\lim_{S \in J} \langle f,g \rangle_S = \langle f,g \rangle ,$$

uniformly on compact subsets of R .

Let ν be a regular Baire measure with compact support $K \subseteq R$ and let $(f,g) \in h^p(R) \times h^q(R)$. Thanks to equation 13,

(14) $$\lim_{S \in J} \int \langle f,g \rangle_S \Big|_a d\nu(a) = \int \langle f,g \rangle \Big|_a d\nu(a)$$

Also, for each $S \in J$ such that $K \subseteq S$,

(15) $$\int_K \langle f,g \rangle_S \Big|_a d\nu(a) = \frac{1}{-2\pi i} \int_{\partial S} fg\Big(\int_K \frac{\delta G_S(a,z)}{\delta G(a_0,z)} d\nu(a)\Big)\delta G(a_0,z) ,$$

since both sides equal

$$\frac{1}{-2\pi i} \int_K \int_{\partial S} fg\delta G_S(a,z)d\nu(a) ,$$

by Fubini's theorem. The joint measurability of the integrand is guaranteed by lemma 6.3.3. We are now able to begin the proof of lemma 2.

Proof: We shall show

(16) $$\int_\gamma \langle h,B \rangle \Big|_a d\psi(a) = [\, h,BP_\gamma \,] ,$$

for each function $h \in h^\infty(R)$. Once equation 16 is established, we may argue as follows: Thanks to theorem 4.4.2,

$$[\, h,BP_\gamma \,] = [\langle h,B \rangle, \int_\gamma q_a d\psi(a) \,] = [\, h, \langle B, \int_\gamma q_a d\psi(a) \rangle] ,$$

for all $h \in h^\infty(R)$. By theorem 4.5.4,

$$BP_\gamma = < B, \int_\gamma q_a d\psi(a) > .$$

From equation 4.4.13, we conclude

$$\phi(B)\phi(P_\gamma) = \phi(B)\phi(\int_\gamma q_a d\psi(a)) = \phi(B)\int_\gamma k_a d\psi(a) \quad \text{a.e.} \quad .$$

Since $\phi(B)$ differs from zero a.e., this will establish lemma 2.

It remains to show that equation 16 actually holds. Let $h \in h^\infty(R)$, and let $\| \ \|_S$ be the $h^2(S)$ norm for each $S \in J$. Let $S \in J$ be such that $\mathrm{Cl}V \subseteq S$. For $T \in J$ such that $\mathrm{Cl}S \subseteq T$,

$$(17) \qquad \int_\gamma < h,B >_T \Big|_a d\psi(a) = < h,BP_{T,\gamma} >_T \Big|_{a_0} ,$$

thanks to equation 15. Also, for such T,

$$(18) \qquad \left| [h,BP_\gamma] - < h,BP_{T,\gamma} >_T \Big|_{a_0} \right|$$

$$\leq \frac{1}{-2\pi i} \int_{\partial S} |< h,BP_\gamma > - < h,BP_{T,\gamma} >_T| \, \delta G(a_0,z)$$

$$\leq A(S) + C(S,T) + D(S,T) .$$

Here,

$$A(S) = \frac{1}{-2\pi i} \int_{\partial S} |< h,BP_\gamma > - hBP_\gamma| \, \delta G(a_0,z)$$

$$C(S,T) = \frac{1}{-2\pi i} \int_{\partial S} |hBP_\gamma - hBP_{T,\gamma}| \, \delta G(a_0,z)$$

$$D(S,T) = \frac{1}{-2\pi i} \int_{\partial S} |hBP_{T,\gamma} - < h,BP_{T,\gamma} >_T| \, \delta G(a_0,z) \quad .$$

Let $S_0 \in J$ be such that $\mathrm{Cl}V \subseteq S_0$. Let $\varepsilon > 0$. Choose $S \in J$ so large that $\mathrm{Cl}S_0 \subseteq S$ and

$$(||h||^2 - ||h||_S^2)^{1/2} < \varepsilon ,$$

where $||\ ||$ denotes the $h^2(R)$ norm. Since the $||\ ||_T$ norms increase with increasing $T \in J$,

$$(||h||_T^2 - ||h||_S^2)^{1/2} < \varepsilon$$

for all $T \in J$ such that $\mathrm{Cl}S \subseteq T$. Thanks to theorem 3.5.2 and inequalities 10 and 12,

$$D(S,T) \leq (||h||_T^2 - ||h||_S^2)^{1/2} (||BP_{T,\gamma}||_T^2 - ||BP_{T,\gamma}||_S^2)^{1/2}$$

$$\leq 2\mu(S_0)\underline{1}(\gamma)(||h||_T^2 - ||h||_S^2)^{1/2} \leq 2\mu(S_0)\underline{1}(\gamma)\varepsilon ,$$

for all such T. Similarly

$$A(S) \leq 2\mu(S_0)\underline{1}(\gamma)\varepsilon .$$

Thanks to equation 11 and the maximum principle, $BP_{T,\gamma} \longrightarrow BP_\gamma$ uniformly on compact subsets of R. Choose $T \in J$ so large that $\mathrm{Cl}S \subseteq T$ and

$$|BP_\gamma - BP_{T,\gamma}| \leq \varepsilon$$

on ∂S. Then

$$C(S,T) \leq ||h||_\infty \varepsilon .$$

Thus

$$(19) \qquad A(S) + C(S,T) + D(S,T) \leq (4\mu(S_0)l(\gamma) + ||h||_\infty)\varepsilon$$

From inequalities 18 and 19, we conclude

$$(20) \qquad \lim_{T \in J} \langle h, BP_{T,\gamma} \rangle_T \Big|_{a_0} = [\, h, BP_\gamma \,] \,.$$

By appealing to equations 14, 17, and 20, we obtain equation 16, which completes the proof of lemma 2.

Chapter 7

The Main Theorem, Counter-examples

7.1. The principal goal of this paper is to classify, for each admissible surface R, the closed subspaces of $H^p(R)$ which are invariant under multiplication by each bounded analytic function on R. We shall attain this goal in theorem 1, which is an extension of the Beurling-Rubel-Shields theorem [3], [35]. The first four sections of this chapter will be devoted to the proof of theorem 1. The remaining four sections will be devoted to a number of counter-examples. In addition, in theorem 7.7.2, we shall give a complete characterization of interior functions for a wide class of plane regions.

A linear subspace of $H^p(R)$ is an <u>invariant subspace</u> if it is an $H^\infty(R)$ submodule of the $H^\infty(R)$ module $H^p(R)$. That is a linear subspace V of $H^p(R)$ is <u>invariant</u> if $fV \subseteq V$ for each $f \in H^\infty(R)$. Although the term "invariant subspace" is firmly embedded in the literature of operator theory, we shall prefer the algebraically more precise terms <u>ideal</u> and $H^\infty(R)$ <u>submodule</u>. Throughout this section, R shall be an admissible surface. The main theorem of this paper is

<u>Theorem</u> 1:

A. Each β closed ideal, V, of $H^\infty(R)$ is of the form

$$V = \{f \in H^\infty(R): \ |f|/I \text{ is bounded}\}, \tag{1}$$

where I is a bounded inner l.a.m. uniquely determined by V. Conversely, each set V of the form 1, where I is again a bounded inner l.a.m., is a β closed ideal of $H^\infty(R)$.

B. Let $1 \leq p < \infty$. Each norm closed $H^\infty(R)$ submodule, V, of $H^p(R)$

is of the form

$$(2) \qquad V = \{f \in H^p(R): \ (|f|/I)^p \ \text{admits a harmonic majorant}\} ,$$

where I is a bounded inner l.a.m. uniquely determined by V. Here, norm closed means closed in $H^p(R)$ norm. Conversely, each set V of the form 2 is a norm closed $H^\infty(R)$ submodule of $H^p(R)$, provided I is a bounded inner l.a.m..

Thanks to the characterizations of the duals of $(h^\infty(R),\beta)$ and $h^p(R)$ given by theorems 4.4.1 and 4.5.4, we have the following immediate corollary to theorem 1:

Corollary 1.1:

A. Statement A of theorem 1 still holds if the phrase "β closed" is replaced by "$\sigma[h^\infty(R),h^1(R)]$ closed" throughout.

B. Statement B of theorem 1 still holds if the phrase "norm closed" is replaced by "$\sigma[h^p(R),h^q(R)]$ closed" throughout.

From theorem 1 and its corollary, we may immediately obtain an analogous theorem for invariant subspaces of automorphic functions on $\Delta(0,1)$. We need only apply the remarks immediately preceding theorem 4.5.4, together with equations 3.3.5, 3.3.6, 3.6.10, and theorem 2.4.2. We shall leave the statement of the automorphic version of theorem 1 to the reader

From the comments immediately preceding definition 2.3.1, and from the discussion at the beginning of section 2.4, we see that if $R = \Delta(0,1)$, then the l.a.m. I in equations 1 and 2 is actually the modulus of an inner function. Thus, part B of theorem 1, for $p = 2$, is a generalization of Beurling's classical invariant subspace theorem [3]. For $p \neq 2$, part B is a generalization of a folk theorem which was first formally enunciated by H. Helson [18, pp. 25-36]. Actually, Helson did more, he in effect classified the closed $H^\infty(\Delta(0,1))$ submodules of $h^p(\Delta(0,1))$ for $1 \leq p < \infty$. Part A of theorem 1 is a generalization of the Rubel-Shields theorem [35], which, as explained in chapter 1, originally motivated this paper. Part A of corollary 1.1 is a generalization of results due to Srinivasan and later to Hasumi and Srinivasan jointly [12].

In all of the results just cited, R was taken to be $\Delta(0,1)$. However, Sarason did theorem 1B and corollary 1.1 for R an annulus, and Voichick obtained theorem 1B and corollary 1.1 for R a non-compact bordered Riemann surface [40,45]. Theorem 1 and corollary 1.1 generalize both of these important results.

7.2. We shall postpone the proof of theorem 7.1.1 and its corollary until section 4. This section and the next will be devoted to a number of preliminary lemmas and remarks. Throughout this section, we shall drop the hypothesis that R is admissible.

First suppose that f and g are arbitrary l.m.m.'s. If $1 \leq p < \infty$, we shall say that f <u>divides</u> g <u>in</u> $h^p(R)$ <u>norm</u> if $(g/f)^p$ admits a harmonic majorant. We shall say that f divides g boundedly, or in $h^\infty(R)$ norm, if g/f is bounded. In either case, we shall write $f|g$ to indicate that f divides g.

<u>Lemma</u> 1: Let f be an l.m.m. of bounded characteristic, and let F be the outer factor of f. Let I be an inner l.m.m.. Then I divides f in $h^p(R)$ norm if and only if I divides the inner factor of f boundedly and

A. F^p has a harmonic majorant, if $1 \leq p < \infty$,

B. F is bounded, if $p = \infty$.

<u>Proof</u>: For simplicity, we shall only consider the case $1 \leq p < \infty$. Let JK be the inner-outer factorization of the l.m.m. f/I. Suppose $I|f$ in $h^p(R)$ norm. Since $(JK)^p$ has a harmonic majorant, J is bounded above by 1 and K^p admits a harmonic majorant, thanks to theorem 2.5.3. But $f = IJK$, IJ is inner, and K is outer. Thus IJ is the inner factor of f, and $K = F$, the outer factor of f. This establishes the only if half of lemma 1. The converse is trivial.

From lemma 1 and theorem 2.5.3, we immediately obtain

Lemma 2: Let f be an l.a.m., and let I be an inner l.m.m..

A. Let $1 \leq p < \infty$. If f^p has a harmonic majorant, then I divides f in $h^p(R)$ norm if and only if I divides the inner factor of f boundedly.

B. If f is bounded, then I divides f boundedly if and only if I divides the inner factor of f boundedly.

In view of lemma 1, we shall dispense with the phrase "in $h^p(R)$ norm" when discussing the divisibility of one inner l.m.m. by another. There is a simple criterion for the divisibility of one inner l.m.m. by another. Let I and J be inner l.m.m.'s. Thanks to theorem 2.5.3 again, $I|J$ if and only if $J/I \leq 1$. Thus $I|J$ if and only if

$$(1) \qquad \ln J \leq \ln I \, .$$

Let $\mathcal{I}$ be a set of inner l.m.m.'s. If J is an inner l.m.m. which divides each element of $\mathcal{I}$, then we shall call J a common inner divisor of $\mathcal{I}$. Suppose that $\mathcal{I}$ admits a common inner divisor J. Thanks to inequality 1,

$$(2) \qquad \ln K \leq \ln J \, ,$$

for each $K \in \mathcal{I}$. Further, $\ln J \in G(R) \oplus I(R)$ by definition 2.3.1. Thus

$$(3) \qquad s = \bigwedge H$$

exists, where

$$H = \{h \in G(R) \oplus I(R) \colon \ln K \leq h \text{ for each } K \in \mathcal{I}\} \, .$$

Of course, H is non-empty by inequality 2, so the existence of s is guaranteed by theorem 2.2.1. In addition, $G(R) \oplus I(R)$ is a band in $SP(R)$ by theorem 2.2.2, so $s \in G(R) \oplus I(R)$. Thus s actually lies in H . Set

$$I = \exp s \tag{4}$$

Thanks to inequality 1, I is a common inner divisor of $\mathcal{I}$, and each common inner divisor of $\mathcal{I}$ divides I . By analogy with the corresponding situation in algebra, we shall call I the greatest common inner divisor of $\mathcal{I}$, or the g.c.i.d. of $\mathcal{I}$.

If $\mathcal{I}$ consists of bounded inner l.a.m.'s, then 1 is a common inner divisor of $\mathcal{I}$. Since 1 divides I, $I \leq 1$. We have just proved

Lemma 3: Each set of inner l.m.m.'s which admits a common inner divisor has a g.c.i.d.. Each set of bounded inner l.a.m.'s admits a g.c.i.d. which is also bounded.

Using the lattice properties of $SP(R)$, we may write each inner l.m.m. as a quotient of bounded inner l.a.m.'s with g.c.i.d. 1. In fact, we can do even better. First, let $h \in SP(R)$. Since $SP(R)$ is a vector lattice,

$$h = h \vee 0 + h \wedge 0 ,$$

[41, p. 207]. Further, if $h = u + v$, where u and $v \in SP(R)$, then $u = h \vee 0$ and $v = h \wedge 0$ if and only if each of the following three conditions are satisfied:

$$u \geq 0 , \quad v \leq 0 , \quad u \wedge (-v) = 0 . \tag{5}$$

Now let f be an l.m.m. of bounded characteristic, and set $h = -\ln f$. We shall call the factorization

(6) $$f = \exp(-(h \vee 0))/\exp(h \wedge 0) ,$$

the minimal factorization of f. Clearly the minimal factorization allows us to write f as a quotient of bounded l.a.m.'s. In addition, the g.c.i.d. of the inner factors of the numerator and the denominator is 1. This result is subsumed under the following lemma:

Lemma 4: Let IF be the inner-outer factorization of f. Let f_1/f_2 be the minimal factorization of f, and let I_iF_i be the inner-outer factorization of f_i for $i = 1,2$, . Then I_1/I_2 and F_1/F_2 are the minimal factorizations of I and F respectively. Furthermore, if J is an inner l.m.m. and if $I_2|I_1J$, then $I_2|J$

Proof: Let $h = -\ln f$, and let P_1 and P_2 be the projection maps of $SP(R)$ onto $G(R) \oplus I(R)$ and $Q(R)$ respectively. Here, P_1 and P_2 are associated with the direct sum decomposition 2.2.2. Thanks to the positivity of P_1 and P_2, we have

$$P_i(h \vee 0) \geq 0 , \quad P_i(h \wedge 0) \leq 0 , \quad P_i(h \vee 0) \wedge (-P_i(h \wedge 0)) = 0 ,$$

for $i = 1,2$. From equation 5, we obtain

$$P_i(h \vee 0) = P_i(h) \vee 0 ,$$
$$P_i(h \wedge 0) = P_i(h) \wedge 0 ,$$

for $i = 1,2$. This, together with equation 6, establishes the first two assertions of lemma 4. As for the third, set $u = -\ln J$, $u_1 = -\ln I_1$, and $u_2 = -\ln I_2$. By inequality 1,

$$u_2 \leq u_1 + u = u_1 + u \vee 0 + u \wedge 0 .$$

Thus, $u_2 - u \wedge 0 \leq u_1 + u \vee 0$, so

$$u_2 - u \wedge 0 = (u_2 - u\wedge 0)\wedge(u_1 + u\wedge 0)$$
$$\le (u_2 + u\vee 0)\wedge(u_1 + u\vee 0) = u_1 \wedge u_2 + u \vee 0 = u \vee 0 .$$

Hence $u_2 \le u$, which implies that $I_2|J$ by inequality 1.

The analogy between the minimal factorization and relatively prime factorizations in algebra should be evident.

7.3. We shall continue to assume, for the time being, that R is merely hyperbolic. Despite the fact that R may not be admissible, it still makes sense to consider subsets of $H^p(R)$ of either the form 7.1.1 or the form 7.1.2. If I is a bounded inner l.a.m., we shall let $H^p(I:R)$ denote the subset of the form 7.1.1 if $p = \infty$, or the subset of the form 7.1.2 if $1 \le p < \infty$. We shall say that a subset V of $H^p(R)$ is quasi-principal if $V = H^p(I:R)$ for some I .

Let I be a bounded inner l.a.m., and consider $H^p(I:R)$ for some $p \in [1,\infty]$. For simplicity, we shall assume $1 \le p < \infty$. For f_1 and $f_2 \in H^p(I:R)$, we have

$$|f_1+f_2|/I \le |f_1|/I + |f_2|/I ,$$

so $|f_1+f_2|/I$ has no poles and thus is subharmonic. Temporarily, let J be the set of all Jordan subregions of R , as in sections 3.1 through 6.2. Then for each $S \in J$ and $a \in S$, an application of the Minkowski inequality shows that

$$\left(\frac{-1}{2\pi}\int_{\partial S}(|f_1+f_2|/I)^p * dG_S(a,z)\right)^{1/p} \le \left(\frac{-1}{2\pi}\int_{\partial S}(|f_1|/I)^p * dG_S(a,z)\right)^{1/p}$$
$$+ \left(\frac{-1}{2\pi}\int_{\partial S}(|f_2|/I)^p * dG_S(a,z)\right)^{1/p} .$$

Passing to the limit over the directed set J , we have

(1) $(\text{L. H. M. }(|f_1+f_2|/I)^p)^{1/p} \le (\text{L. H. M. }(|f_1|/I)^p)^{1/p} + (\text{L. H. M. }(|f_2|/I)^p)^{1/p}$

In particular, $(|f_1+f_2|/I)^p$ admits a harmonic majorant, so $f_1 + f_2 \in H^p(I:R.)$.

Clearly $hf_1 \in H^p(I:R)$ for each $h \in H^\infty(R)$. We have almost proved

Lemma 1: For $1 \le p < \infty$, $H^p(I:R)$ is a norm closed $H^\infty(R)$ submodule of $H^p(R)$. For $p = \infty$, $H^p(I:R)$ is a β closed ideal in $H^\infty(R)$.

Proof: We only need to show that $H^p(I:R)$ is closed. For the sake of purity of method, we shall not use the universal covering surface of R, even though it would shorten the proof slightly. First, suppose $1 \le p < \infty$. Let f_n, $n = 1,2,\cdots$, be a sequence in $H^p(I:R)$ converging to f in $h^p(R)$ norm. The sequence f_n, $n = 1,2,\cdots$, is bounded in $h^p(R)$ norm, say by k. Let J_nF_n be the inner-outer factorization of the l.a.m. $|f_n|/I$. As in the proof of lemma 7.2.1, F_n is the outer factor of f_n, and

$$\text{L. H. M. } |f_n|^p = \text{L. H. M. } |F_n|^p = \text{L. H. M. } (|f_n|/I)^p .$$

Thus

$$\text{L. H. M. } (|f_n|/I)^p\Big|_{a_0} \le k \tag{2}$$

for $n = 1,2,\cdots$.

In addition, the sequence $f_n \longrightarrow f$ uniformly on compact subsets of R. If I has a zero of order c at some point w, each f_n has a zero of order c or larger at w, thanks to inequality 7.2.1. From the uniform convergence of the sequence f_n, $n = 1,2,\cdots$, on compact subsets, we see that f also has a zero of order c or larger at w. Thus $|f|/I$ is an l.a.m.. Let

$$J_0 = \{S \in J\text{: } I \text{ has no zeros on } \partial S\} .$$

A simple cardinality argument shows that J_0 is cofinal in J. Let $\varepsilon > 0$, and let $S \in J_0$. By uniform convergence again, there exists N such that $n \ge N$ implies

$$|f_n - f|/I \le \varepsilon ,$$

on ∂S . As in the derivation of inequality 1,

$$(\text{L. H. M. } (|f|/I|_S)^p)^{1/p}$$

$$\leq \varepsilon + (\text{L. H. M. } (|f_n|/I|_S)^p)^{1/p} .$$

An application of inequality 2 shows that

$$(\text{L. H. M. } (|f|/I|_S)^p)^{1/p}|_{a_0} \leq \varepsilon + k ,$$

for each $S \in J_0$. Passing to the limit over the directed set J_0 and applying the Harnack monotone convergence theorem, we obtain the existence of a harmonic majorant for $(|f|/I)^p$. Thus $f \in H^p(I:R)$, so $H^p(I:R)$ is norm closed.

Suppose $p = \infty$. Since $H^p(I:R)$ is a subspace, it suffices to show that $H^p(I:R)$ is β sequentially closed, thanks to theorem 4.5.2. The rest of the proof is similar to the case where $1 \leq p < \infty$.

We now adjoin the hypothesis that R is an admissible surface admitting the system of multipliers 5.2.1.

Lemma 2: Let I and J be bounded inner l.a.m.'s, and suppose $I \neq J$. Then $H^p(I:R) \neq H^p(J:R)$.

Proof: Choose ε from the system of multipliers 5.2.1 so that $J\varepsilon$ has character 1. Let $f \in H^\infty(R)$ be a global branch of $J\varepsilon$. Note that J is the inner factor of f , and that $f \in H^p(J:R)$. Suppose that $H^p(I:R) = H^p(J:R)$. Thanks to lemma 7.2.1, the fact that $f \in H^p(I:R)$ implies that $I|J$. From inequality 7.2.1, we conclude that $J \leq I$. Similarly $J \leq I$, so $I = J$, which is impossible.

7.4. We are now ready to begin the proof of the main theorem of this paper, theorem 7.1.1. Our proof is modeled on Rudin's proof of the Beurling-Rudin characterization of the closed ideals of the algebra of continuous functions on $\mathrm{Cl}\Delta(0,1)$ which are analytic on $\Delta(0,1)$ (cf. [22, pp. 85-87]). In many respects, our proof also follows Voichick's proof of an analogue of theorem 7.1.1 for finite open Riemann surfaces [45]. We are particularly indebted to Voichick's paper for the idea of using the theorem of A. H. Read at a crucial point in the proof [32]. Of course, we shall actually use our extension of Read's theorem, theorem 6.3.1, rather than Read's original theorem.

Throughout this section, R shall be an admissible surface admitting the system of multipliers 5.2.1. Also, g and B shall be the functions given in formulas 6.2.1 and 6.2.2 respectively.

Proof of theorem 7.1.1: We shall prove parts A and B together. Let $V \subseteq H^p(R)$. If $p = \infty$, suppose V is a β closed ideal. If $1 \leq p < \infty$, suppose V is a norm closed $H^\infty(R)$ submodule. In either case, let I be the g.c.i.d. of the set of inner factors of the set $|V| = \{|f|: f \in V\}$. Thanks to theorem 2.5.3, the inner factor of each element of $|V|$ is bounded, which, together with lemma 7.2.3, guarantees the existence and boundedness of I. An application of lemma 7.2.2 shows that

$$V \subseteq H^p(I:R)$$

From lemma 7.3.1, we see that $H^p(I:R)$ is a closed subspace. Thus, from the Hahn-Banach theorem and the duality results contained in theorems 4.4.1 and 4.5.4, to show that $V = H^p(I:R)$, it suffices to show the following: If $s \in h^q(R)$ and if

$$\int \phi(f)\phi(s)d\omega = 0 , \tag{1}$$

for each $f \in V$, then

(2) $$\int \phi(f)\phi(s)d\omega = 0 ,$$

for each $f \in H^p(I:R)$.

Let s lie in $h^q(R)$ and satisfy equation 1 . Let $f \in V$. Also, let h be a meromorphic function on R such that $g|h|$ is bounded. Then $Bh \in H^\infty(R)$. Since V is a submodule, $Bhf \in V$. Applying equation 1 to Bhf , we conclude

$$\int \phi(h)\phi(Bf)\phi(s)d\omega = 0 .$$

Thanks to theorem 6.3.1, our generalization of Read's theorem, we have

(3) $$\phi(Bf)\phi(s) = \phi(H_f)$$

for some function $H_f \in H^1(R)$. Furthermore, the function

$$\psi_f = H_f/Bf$$

is meromorphic, of bounded characteristic, and

(4) $$\phi(\psi_f) = \phi(s)$$

a.e. $d\omega$. We have used lemma 6.1.2 to obtain equation 4 from equation 3. Because the map 6.1.5 is linear, another application of lemma 6.1.2 shows that ψ_f is independent of f , provided $f \in V$. Set $\psi = \psi_f$. To summarize, ψ is meromorphic, of bounded characteristic on R , and

(5) $$\phi(\psi) = \phi(s) \quad \text{a.e. } d\omega ,$$
$$Bf\psi \in H^1(R) ,$$

for each $f \in V$. We shall show that a variant of equation 5 holds for f lying in $H^p(I:R)$. We shall then apply our generalized Cauchy integral formula, that is theorem 6.2.1, to conclude that equation 2 holds.

Let

$$|\psi| = I_1F_1/I_2F_2 \ ,$$

be the minimal factorization of $|\psi|$. Here I_1 and I_2 are bounded inner l.a.m.'s, and F_1 and F_2 are bounded outer l.a.m.'s. The required variant of equation 5 is

Lemma 1: For each $f \in H^p(I:R)$, $gF_2|f||\psi|$ admits a harmonic majorant.

Proof: For each $f \in V$, the inner factor of $Bf\psi$ is

$$gJI_1/I_2 \ , \tag{6}$$

where J is the inner factor of $|f|$. Thanks to equation 5 and theorem 2.5.3, the l.a.m. 6 is bounded. From lemma 7.2.4, we see that $I_2|gJ$, so I_2/g divides J . Since I is the g.c.i.d of the inner factors of $|V|$, I_2/g divides I . Thus $I_2|gI$, so for each $f \in H^p(I:R)$, I_2 divides the inner factor of $g|f|$. Thanks to lemma 7.2.2, $I_2 \big| g|f|$ in $h^p(R)$ norm. Thus $gF_2|f||\psi|$ admits a harmonic majorant for each $f \in H^p(I:R)$.

Proof of theorem 7.1.1 continued: Let $f \in H^p(I:R)$. Choose ε from the system of multipliers 5.2.1 so that εF_2 has a global branch $u \in H^\infty(R)$. For each $h \in H^\infty(R)$, $g|huf\psi|$ admits a harmonic majorant, thanks to lemma 1. By applying theorem 6.2.1, that is our generalized Cauchy integral theorem, we obtain

$$huf\psi\Big|_{a_0} = \int \phi(hu)\phi(f)\phi(\psi)d\omega \ .$$

However, u is β exterior by theorem 5.2.1, since εF_2 is outer. Also,

(7) $$\phi(f)\phi(\psi) = \phi(f)\phi(s) \quad \text{a.e.} ,$$

so $\phi(f)\phi(\psi) \in L^1(d\omega)$. Choosing a net h_α, $\alpha \in A$, such that $h_\alpha u \longrightarrow 1$ β , and applying corollary 4.5.4.2, we obtain

(8) $$f\psi\Big|_{a_0} = \int \phi(f)\phi(\psi)d\omega ,$$

for each $f \in H^p(I:R)$. Thanks to equations 7 and 8, we need only prove the following lemma to complete the proof that equation 2 holds, and thus to complete the proof that $V = H^p(I:R)$:

Lemma 2: For each $f \in H^p(I:R)$,

(9) $$f\psi\Big|_{a_0} = 0 .$$

Proof: We shall divide the proof into two cases. First, suppose V contains a function f_0 such that $f_0(a_0) \neq 0$. Then equations 1 and 8, with f_0 substituted for f , show that $\psi(a_0) = 0$.

As for the contrary case, suppose each $f \in V$ has a zero at a_0 . Let

$$n = \inf\{c(f): \ f \in V\} ,$$

where $c(f)$ is the order of the zero of f at a_0 . Note that there exists a function $f_0 \in V$ such that $n = c(f_0)$. From equations 1 and 8, we conclude that $f_0\psi\Big|_{a_0} = 0$. Thus ψ has a pole of order $\leq n - 1$ at a_0 . But equation 2.2.1, together with equations 7.2.3 and 7.2.4, show that I has a zero of order n at a_0 . Thus equation 9 holds for all $f \in H^p(I:R)$.

The remaining assertions in theorem 7.1.1 are contained in lemmas 7.3.1 and 7.3.2. Actually, we have proved more than is asserted by theorem 7.1.1. We have also proved

Theorem 1: The bounded inner l.a.m. I occuring in equations 7.1.1 and 7.1.2 is the g.c.i.d. of the set of inner factors of $|V| = \{|f|: f \in V\}$.

7.5. In the remainder of this chapter, we shall, unless otherwise stated, drop the hypothesis that R is admissible. We shall also use the symbol $\hat{\mathbb{C}}$ to denote the extended complex plane. In this section, we shall give the first of the promised counter-examples. We shall construct a hyperbolic plane region R for which theorem 7.1.1A fails utterly. In fact, we shall produce a β closed maximal ideal in $H^\infty(R)$ which is not quasi-principal! The construction is based on Rudin's construction of a plane region R and a β closed maximal ideal in $H^\infty(R)$ which is not the ideal of all functions vanishing at some point of R [39]. The construction is also based on some basic facts about logarithmic capacities. We quote, with slight notational changes, the relevant theorem, due to Norbert Wiener, from Tsuji [44, p. 104]:

Theorem 1: Let E be a bounded closed set of positive capacity in the z plane and b a point on the outer boundary of E. Let $0 < \lambda < 1$ and E_n be the part of E contained in $\lambda^{n+1} \leq |z-b| \leq \lambda^n$. Then b is a regular point for the Dirichlet problem for D_∞ when and only when

$$(1) \qquad \sum_{n=1}^{\infty} \frac{n}{\ln(1/\gamma_n)} = \infty ,$$

where $\gamma_n = \gamma(E_n)$ = the capacity of E_n.

Here D_∞ is the unbounded component of $\hat{\mathbb{C}} \sim E$, and the outer boundary of E is just ∂D_∞. In addition to the theorem just quoted, we need to know that $\gamma(\mathrm{Cl}\Delta(a,r)) = r$. The promised counter-example is

Counter-example 1: Let

$$(2) \qquad a_k = -i2^{-k}(3/4) , \quad \rho_k = 4^{-k}(1/4) ,$$

for $k = 1,2,\cdots$. Let $R_0 = \hat{\mathbb{C}} \sim (\{0\} \cup \bigcup\{\mathrm{Cl}\Delta(a_k,\rho_k): k=1,2,\ldots\})$ and

(3) $$R = \Delta(0,1) \cap R_0$$

Then R is a hyperbolic region in the extended plane. Furthermore, if

(4) $$I = \{f \in H^\infty(R) : \lim_{\substack{z\to 0\\ Imz>0}} f(z) = 0\} ,$$

then I is a β closed maximal ideal in $H^\infty(R)$, but I is not quasi-principal.

Proof: Clearly R is open and connected, and

$$\partial R = \{0\} \cup \bigcup\{\partial\Delta(a_k,\rho_k): k = 1,2,\cdots\} \cup \partial\Delta(0,1) .$$

Also, R obviously admits non-constant positive super-harmonic functions. Furthermore,

(5) $$\sum_{k=1}^{\infty}\rho_k/|a_k| = \sum_{k=1}^{\infty} (1/3)\ 2^{-k} < \infty .$$

Let

$$I_0 = \{f \in H^\infty(R_0) : \lim_{\substack{z\to 0\\ Imz>0}} f(z) = 0\} .$$

The convergence of the left hand series in equation 5 is exactly the condition used by Rudin to show that the ideal I_0 is actually a β closed maximal ideal [39]. His proof carries over immediately to the ideal I. Thus we need only show that I is not quasi-principal. For this, we need the following lemma:

Lemma 1: The function $z \longrightarrow \log|z|$, $z \in R$, is quasi-bounded.

Once lemma 1 has been established, we may quickly complete the proof of counter-example 1. The function $z \longrightarrow z$, $z \in R$, lies in I , and the inner factor of $|z|$, $z \in R$, is 1 . If I were of the form $H^\infty(I:R)$ for some bounded inner l.a.m. I , then I would have to divide 1 by lemma 7.2.1. Also 1 divides I , so I = 1 by equation 7.2.1. But, $H^\infty(I:R) = H^\infty(R) \neq I$, which is impossible.

Theorem 7.1.1 cannot be changed to be true for R simply by dropping the requirement that I be bounded in equation 7.1.1. A mild variant of the preceding proof still shows that if $I = H^\infty(I:R)$, then $H^\infty(I:R) = H^\infty(R) \neq I$, which is impossible.

Proof of lemma 1: By contrast with the preceding portions of this paper, throughout this section $d\omega_a$ shall represent harmonic measure at a on ∂R rather than on the maximal ideal space of $h^\infty(R)$, for each $a \in R$ (cf. [5, p. 28]). To prove lemma 1, we shall show that $\ln|z|$ is integrable with respect to harmonic measure on ∂R, and that

$$\ln|z| = \int_{\partial R} \ln|s|\,d\omega_z(s) , \tag{6}$$

for each $z \in R$. Once equation 6 has been established, an argument similar to the comments following equation 2.3.1 immediately shows that the function $\ln|z|$, $z \in R$, is quasi-bounded.

To establish equation 6, we first must show that each point of ∂R is a regular point for the solution to the Dirichlet problem for R. First, consider the point 0. Note that

$$\sum_{n=1}^{\infty} \frac{n}{\ln(1/\rho_n)} = \sum_{n=1}^{\infty} \frac{n}{(n+1)\ln 4} = \infty .$$

Set $E = \hat{\mathbb{C}} \sim R_0$ and note that E is closed, bounded, and of positive capacity. In the notation of theorem 1,

$$E_n = E \cap \{z: \quad (1/2)^{n+1} \leq |z| \leq (1/2)^n\} = \Delta(a_n, \rho_n) ,$$

for $n = 1,2,\cdots$, where $\lambda = 1/2$. Thus

$$\sum_{n=1}^{\infty} \frac{n}{\ln(1/\gamma_n)} \doteq \sum_{n=1}^{\infty} \frac{n}{\ln(1/\rho_n)} = \infty .$$

An application of theorem 1 shows that 0 is a regular point of R_0. By a standard argument, 0 is also a regular point of R [5, p. 26]. Since $\partial R \sim \{0\}$ is a union of continua, each point of $\partial R \sim \{0\}$ is also a regular point.

For each $a \in R$, consider the function

$$h(a,z) = \int_{\partial R} \ln |s-a| d\omega_z(s) \quad , \ z \in R$$

$$h(a,z) = \ln |z-a| \quad , \quad z \in \partial R.$$

Thanks to regularity, the function $z \longrightarrow h(a,z)$ is continuous on $\mathrm{Cl}R$ for each $a \in R$. An elementary argument shows that

$$G(a,z) = -\ln |z-a| + h(a,z) \ ,$$

on $R \times R$ (cf. [16, pp. 96-97]). Since both $G(a,z)$ and $\ln |z-a|$ are symmetric on $R \times R$, so is $h(a,z)$. Thus

(7) $$\lim_{\substack{a \to 0 \\ a \in R}} h(a,z) = \lim_{\substack{a \to 0 \\ a \in R}} h(z,a) = \ln |z| \ ,$$

for each $z \in R$. Let t_n, $n = 1,2,\cdots$, be a decreasing sequence in $(0,1)$ with limit 0. If $s \in \partial R \sim \{0\}$, then $\ln |s-it_n| \downarrow \ln |s|$ as $n \to \infty$. Thanks to the Phragmén-Lindelöf maximum principle, the set $\{0\}$ has harmonic measure zero, so $\ln |s-it_n| \downarrow \ln |s|$ a.e. on ∂R with respect to harmonic measure. By equation 7 and the Lebesgue monotone convergence theorem, $\ln |s|$ is integrable on ∂R, and

$$\ln |z| = \lim_{n\to\infty} h(it_n, z) = \lim_{n\to\infty} \int_{\partial R} \ln |s-it_n| d\omega_z(s) = \int_{\partial R} \ln |s| d\omega_z(s) \ ,$$

for each $z \in R$. This completes the proof of lemma 1.

Recently Kennel has generalized counter-example 1 significantly. He has given necessary and sufficient conditions on a class of plane regions for the function z to belong to a proper β closed maximal ideal in $H^\infty(R)$, necessary and sufficient conditions for $\mathrm{Cl} z \cdot H^\infty(R)$ to equal this maximal ideal, and necessary and sufficient conditions under which $\text{Clz}\cdot H^{\infty}(R)$ fails to be principal [24].

7.6. Let V be an $H^{\infty}(R)$ submodule of $H^p(R)$. We shall say that V is <u>principal</u> if

$$V = \psi H^p(R) , \tag{1}$$

where ψ is some function in $H^{\infty}(R)$. If V is of the form 1, we shall call ψ a <u>generator</u> of V. For a wide class of surfaces, including the admissible surfaces, each closed principal submodule turns out to be quasi-principal. The question thus arises, can theorem 7.1.1 be improved to assert that each β closed ideal of $H^{\infty}(R)$ (respectively each norm closed submodule of $H^p(R)$) is principal? The answer turns out to be no! To completely produce an example of an admissible surface R and a quasi-principal submodule of $H^p(R)$ which is not principal, we would have to introduce some of the extensive machinery contained in [26], but superceded in this paper by chapters 3 and 4. We shall, therefore, merely sketch much of the construction. Pending the publication of a paper devoted to this construction, the interested reader may consult [26, pp. 217-427] for details. In this section, we shall prove those results pertaining to principal submodules which can be obtained without the introduction of extensive machinery. In the remaining two sections, we shall sketch the proofs of those results which depend on the machinery of [26].

First, let us broaden the notions of interior and exterior functions introduced in section 1.2.

<u>Definition</u> 1: Let V be a submodule of the form 1. Suppose $\psi \not\equiv 0$. If $p = \infty$ and V is β closed, we shall say that ψ is β <u>interior</u>. If $1 \leq p \leq \infty$ and V is norm closed, we shall say that ψ is p <u>interior</u>.

Suppose W is a submodule of the form

$$W = fH^{\infty}(R) , \tag{2}$$

where $f \in H^p(R)$. If $p = \infty$ and the β closure of W is $H^\infty(R)$, we shall say that f is β exterior. If $1 \leq p \leq \infty$ and the norm closure of W is $H^p(R)$, we shall say that f is p exterior.

If R is admissible, then thanks to theorem 7.1.1, we know that $f \in H^p(R)$ is p exterior if and only if $|f|$ is outer. Similarly, $f \in H^\infty(R)$ is β exterior if and only if $|f|$ is outer. On the other hand, if R is the region 7.5.3, then, thanks to lemma 7.5.1, the l.a.m. $|z|$, $z \in R$, is outer, but the function z , $z \in R$, is not β exterior because z lies in the β closed maximal ideal 7.5.4. By contrast, for arbitrary hyperbolic surfaces R , we have the following characterization of interior functions:

Theorem 1: Let $\psi \in H^\infty(R)$. Then ψ is p interior if and only if there exist constants k and K such that $0 < k < K < \infty$ and

$$k||f||_p \leq ||\psi f||_p \leq K||f||_p , \tag{3}$$

for all $f \in H^p(R)$. Furthermore, ψ is β interior if and only if ψ is ∞ interior.

Here, $||\ \ ||_p$ denotes the $h^p(R)$ norm. The last assertion of theorem 1 has already been given by Rubel and Shields for domains in C^n [36, Theorem 3]. Let $p = \infty$. As Rubel and Shields observed, if the ideal I is β closed, it is also norm closed, because the β topology is weaker than the norm topology. Thus ψ is ∞ interior if it is β interior. Suppose that $1 \leq p \leq \infty$ and that ψ is p interior. Following Rubel and Shields, we consider the multiplication operator m_ψ: $H^p(R) \longrightarrow \psi H^p(R)$, defined by $m_\psi(f) = \psi f$, $f \in H^p(R)$. Thanks to p interiority, $\psi H^p(R)$ is a Banach space. The operator m_ψ is 1-1, so an application of the open mapping theorem (or the closed graph theorem) establishes 3. To prove the rest of theorem 1, we depart from the treatment given by Rubel and Shields.

Proof: We need only show that the truth of 3 implies ψ is p interior, if $1 \leq p < \infty$, or β interior if $p = \infty$. This, however, follows as in the proof of lemma 7.3.1.

Now let us show that the notion of a quasi-principal submodule actually is a generalization of the notion of a principal submodule for admissible surfaces.

Theorem 2: Let R be an admissible surface. Let V be of the form 1, and let I be the inner factor of $|\psi|$. Then

A. If $p = \infty$ and V is β closed, then $V = H^{\infty}(I:R)$,

B. If $1 \leq p \leq \infty$ and V is norm closed, then $V = H^{p}(I:R)$.

Proof: In view of theorem 1, A and the $p = \infty$ case of B are equivalent. We shall prove A and the $1 \leq p < \infty$ case of B together. Clearly I is the g.c.i.d. of the inner factors of the elements of

$$\{|f|: f \in \psi H^{p}(R)\} .$$

Thanks to theorems 7.1.1 and 7.4.1, $V = H^{p}(I:R)$.

7.7. It is ironic that the machinery developed in [26], although much less powerful than the machinery developed in chapters 3 and 4 of this paper when it comes to proving theorems, nevertheless seems to be much more convenient for constructing examples. In this section, we shall sketch that part of the machinery of [26] necessary for the production of the remaining counter-examples. Throughout this section, unless otherwise stated, $R \cup \Gamma$ shall be a bordered Riemann surface with interior R and the border Γ. Imbed $R \cup \Gamma$ in its Schottky double $\hat{R}$. Then R is a hyperbolic subregion of R, and $\partial_{\hat{R}} R = \Gamma$. As usual, $d\omega_a$ shall represent harmonic measure at a on $\partial_{\hat{R}} R$, for each $a \in R$. Terms such as a.e. and integrable, unless otherwise stated, shall be taken with respect to harmonic measure.

In [26], we showed, if there was ever any doubt, that

(1) $$d\omega_a(z) = \frac{-1}{2\pi i}\,\delta G(a,z) ,$$

on Γ for each $a \in R$. Consequently, a subset A of Γ is a set of harmonic measure 0 if and only if, for each parametric semidisk (V,ψ) in $R \cup \Gamma$, $\psi(V \cap A)$ is a set of Lebesgue measure 0 on $\psi(V \cap \Gamma) = (-1,1)$. Let $f \in h^1(R)$. Using the standard mapping τ of $\Delta(0,1)$ onto $\Delta(0,1) \cap \{Imz>0\}$ (c.f. [20, p. 307]), and applying standard results for $\Delta(0,1)$ to $f\circ\psi^{-1}\circ\tau$ for each parametric semidisk (V,ψ) in $R \cup \Gamma$, we obtain the existence a.e. of the non-tangential, or <u>sectoral, boundary limit function</u> for f. As canonical notation, let f^* be the sectoral boundary limit function for f. It is easy to show that f^* is harmonically measurable. Similarly, if F is an l.m.m. of bounded characteristic, by considering a branch f_V of F on $V \cap R$ for each parametric semidisk (V,ψ), we obtain the existence a.e. on Γ of the sectoral boundary limit function F^*. If F is inner, it is not hard to show that $F^* = 1$ a.e. on Γ. If F is a meromorphic function of bounded characteristic, F^* also exists a.e. on Γ, and whether or not F is an l.m.m. or a meromorphic function of bounded characteristic, F^* is measurable.

Most of this may be done using the Martin boundary and fine limits, but the point is that it can also be done efficiently by elementary means for bordered surfaces. Our approach is similar, but not identical, to the one taken by Royden for compact bordered Riemann surfaces [33]. We certainly don't claim that the machinery just sketched is original. On the contrary, these results are folk theorems which are usually assumed to some extent by mathematicians requiring information about sectoral boundary values on bordered Riemann surfaces (c.f. [7], [8], [51]).

We shall say that $R \cup \Gamma$ is <u>harmonically complete</u> if the ideal boundary of $R \cup \Gamma$ has harmonic measure 0. For harmonically complete bordered surfaces, we have a theorem analogous to theorem 4.2.1 and equation 4.2.3.

<u>Theorem</u> 1: Let $R \cup \Gamma$ be harmonically complete. Then the map

$$f \longrightarrow f^*, \quad f \in h^p(R), \tag{2}$$

is a linear, order preserving isometry of $h^p(R)$ onto $L^p(d\omega_{a_0})$. Furthermore, for each $f \in h^p(R)$,

$$f(a) = \int_\Gamma f^* d\omega_a, \tag{3}$$

for each $a \in R$.

The key to the proof of theorem 1 is the following lemma, which holds whether or not $R \cup \Gamma$ is harmonically complete:

Lemma 1: Let $f \in HB(R)$, and suppose $f^* \geq 0$ a.e. on Γ. Then for each $w \in \Gamma$,

$$\liminf_{\substack{z \to w \\ z \in R}} f(z) \geq 0 .$$

Lemma 1 may be proved by applying results from $\Delta(0,1)$ to $f \circ \psi^{-1}$ for each parametric semidisk (V,ψ) in $R \cup \Gamma$.

We shall now define the class of plane regions to which the remaining results of this chapter will apply.

Definition 1: A region R in the extended complex plane is simple if

$$R = \hat{\mathbb{C}} \sim \bigcup \{A_i : i \in I\}, \tag{4}$$

and if the following conditions are satisfied: The index set I is either finite or countable. Each A_i is a non-void connected compact set such that $\hat{\mathbb{C}} \sim A_i$ is also connected. For $i \neq j$, $A_i \cap A_j = \emptyset$. At least one A_i is a continuum. Finally, the set E of cluster points of the A_i's intersects only finitely many of the A_i's.

We shall call E the <u>exceptional set</u> of R. Since isolated points of $\hat{\mathbb{C}} \sim R$ are removable singularities for $h^1(R)$ functions, we shall suppose without loss of generality that each isolated A_i is a continuum. It is not difficult to show that A_i, $i \in I$, is a univalent enumeration of the components of $\hat{\mathbb{C}} \sim R$. Clearly, each Blaschke region is a simple region, as is each Jordan subregion of $\hat{\mathbb{C}}$ (c.f. definition 5.3.2 f.f.). Thus there are lots of examples of admissible surfaces which are also simple regions. In the opposite direction, the region 7.5.3 is simple.

Each simple region may be imbedded in a bordered Riemann surface in the following way: For each isolated A_i, map $\hat{\mathbb{C}} \sim A_i$ onto $\Delta(0,1)$ via the Riemann mapping theorem. Call the resulting map τ_i and let $\Gamma_i = \partial\Delta(0,1) \times \{\tau_i\}$. For each A_i which is continuum intersecting E, let τ_i be as above, let E_i be the intersection of the cluster set of the $\tau_i(A_j)$, $j \neq i$, with $\partial\Delta(0,1)$, and set

$$\Gamma_i = (\partial\Delta(0,1) \sim E_i) \times \{\tau_i\} .$$

Let $\Gamma = \bigcup\{\Gamma_i : A_i$ is a continuum$\}$. Endow $R \cup \Gamma$ with the obvious topology and conformal structure. Then $R \cup \Gamma$ is a bordered Riemann surface with interior R and border Γ, and the topology and conformal structure induced on R by $\hat{\mathbb{C}}$ agree with the ones induced on R by $R \cup \Gamma$. We shall call Γ the <u>canonical border</u> of R.

Suppose each A_i which is a continuum is a closed, simply connected Jordan subregion of $\hat{\mathbb{C}}$ bounded by an analytic curve. Then $R \cup \Gamma$ is conformally equivalent to $(\mathrm{Cl}R) \sim E$, where E is the exceptional set of R. We shall call $(\mathrm{Cl}R) \sim E$, or equivalently $R \cup \Gamma$, a <u>bordered simple region</u>. If each A_i is either a point or a closed disk, then $R \cup \Gamma$ is clearly a bordered simple region. In particular, if R is the region 7.5.3, then $R \cup \Gamma = (\mathrm{Cl}R) \sim \{0\}$ is a bordered simple region. As we showed by an application of the Phragémen-Lindelöf principle in the proof of lemma 7.5.1, $R \cup \Gamma$ is also harmonically complete. Similarly, if $R \cup \Gamma$ is any bordered simple region for which the exceptional set E is either finite or countable, then $R \cup \Gamma$ is harmonically complete.

Throughout the remainder of this chapter, R shall always be a simple region, Γ shall always be the canonical border of R, and $R \cup \Gamma$ shall be harmonically complete. For such R, we may completely characterize interior functions in terms of the inner-outer factorization of the corresponding l.a.m.'s.

Theorem 2: Let $\psi \in H^\infty(R)$. Then

A. ψ is p interior for some $p \in [1,\infty]$ if and only if ψ is both p interior for all $p \in [1,\infty]$ and β interior,

B. ψ is interior if and only if there exist constants k and K such that $0 < k < K$ and

(5) $$k \le |\psi^*| \le K \quad \text{a.e. on } \Gamma .$$

C. Let IF be the inner-outer factorization of $|\psi|$. Then ψ is interior if and only if both F and $1/F$ are bounded l.a.m.'s.

In view of assertion A, we shall usually omit the p or the β in front of the term "interior function" throughout the rest of this chapter. Assertions B and C use this new terminology. Assertion C is a generalization of a theorem due to Rubel and Shields for $\Delta(0,1)$ [35]. For if $R = \Delta(0,1)$, we may actually take IF to be the inner-outer factorization of ψ, so that assertion C then states that ψ is interior if and only if its outer factor is a unit in $H^\infty(R)$.

Sketch of the proof of theorem 2: Clearly, it suffices to show that ψ is p interior for some $p \in [1,\infty]$ if and only if 5 holds. (To obtain C from this, remember that $I^* = 1$ a.e., $|\psi^*| = I^*F^* = F^*$ a.e., and $(\ln F)^* = \ln(F^*)$ a.e.. Since $\ln F \in h^1(R)$, an application of theorem 1 shows that $\ln k \le \ln F \le \ln K$ if equation 5 holds.) The key to the proof is the characterization of interior functions given by theorem 7.6.1. First, suppose ψ satisfies equation 5. Let $f \in H^p(R)$. Then $k|f^*| \le |(\psi f)^*| \le K|f^*|$. Thanks to theorem 1, equation

7.6.3 holds, so ψ is p interior. Conversely, suppose that ψ is p interior for some $p \in [1,\infty]$. Without loss of generality, $||\psi||_\infty \leq 1$. By interiority, equation 7.6.3 holds. Suppose $|\psi^*| < k$ on a set of positive measure. By countability, $|\psi^*| < \eta k$ on a set W of positive measure on some Γ_i, where $0 < \eta < 1$. Thanks to the construction of Γ, we may assume $A_i = \hat{\mathbb{C}} \sim \Delta(0,1)$, $R \subseteq \Delta(0,1)$, and $\Gamma_i \subseteq \partial\Delta(0,1)$. Then W is a set of positive $d\theta$ measure on $\partial\Delta(0,1)$. It is easy to construct a function $f \in H^p(\Delta(0,1))$ such that equation 7.6.3 is violated for $f|_R$. We shall only consider the case $p = \infty$, although the $1 \leq p < \infty$ case is similar. First, note that there exists a point $t \in \partial\Delta(0,1)$ and $\varepsilon > 0$, such that $N = \Delta(t,\varepsilon) \cap W$ has positive Lebesgue measure, and $\mathrm{Cl}\Delta(t,2\varepsilon) \cap A_j = \emptyset$ for $j \neq i$. Let u be the characteristic function of $\partial\Delta(0,1) \sim N$, and set

$$(6) \qquad g(z) = \exp\left(\frac{1}{2\pi}\int_0^{2\pi} \frac{e^{i\theta}+z}{e^{i\theta}-z} \ln(\eta k)u(e^{i\theta})d\theta\right) .$$

Clearly $g \in H^\infty\Delta(0,1)$, $|g^*| = \eta k$ a.e. on $\partial\Delta(0,1) \sim N$, $|g^*| = 1$ a.e. on N, and $||g||_\infty = 1$. Since u is continuous on $\partial\Delta(0,1) \sim \mathrm{Cl}\Delta(t,\varepsilon)$, the function $|g|$ may be extended continuously to $\mathrm{Cl}\Delta(0,1) \sim \mathrm{Cl}\Delta(t,\varepsilon)$. Denote this extension by s. Now s is subharmonic on $\Delta(0,1)$, continuous on $\mathrm{Cl}\Delta(0,1) \sim \Delta(t,2\varepsilon)$, $s = \eta k$ on $\partial\Delta(0,1) \sim \mathrm{Cl}\Delta(t,\varepsilon)$, and $0 \leq s \leq 1$. By the maximum principle, $s < 1$ on $\mathrm{Cl}\Delta(0,1) \sim \Delta(t,2\varepsilon)$.

Since $\mathrm{Cl}\Delta(0,1) \sim \Delta(t,2\varepsilon)$ is compact, there exists $0 < \zeta < 1$ such that $s < \zeta$ on $\mathrm{Cl}\Delta(0,1) \sim \Delta(t,2\varepsilon)$. Choose n so large that $\zeta^n < \eta k$. Set $f = g^n$. Then $|f| < \eta k$ on $\Delta(0,1) \sim \mathrm{Cl}\Delta(t,2\varepsilon)$, so $|f| < \eta k$ on ∂A_j for each $j \neq i$. Consequently,

$$\limsup_{\substack{z \to w \\ z \in R}} |f(z)| < \eta k ,$$

for each $w \in \Gamma_j$, for $j \neq i$, and thus $|(f|_R)^*| < \eta k$ a.e. on $\Gamma \sim N$. On N, $|(f|_R)^*| = 1$ a.e.. But $|\psi^*| \leq 1$ a.e. on Γ, and $|\psi^*| \leq \eta k$ on N. Thus

$$|(\psi f)^*| = |\psi^*||(f|_R)^*| \le \eta k ,$$

a.e. on Γ . An application of theorem 1 shows that

$$||\psi f||_\infty < \eta k||f|_R||_\infty ,$$

which violates equation 7.6.3.

The reader should note that the proof of theorem 2 has been given quite completely, at least for the $p = \infty$ case, and hence for the β interior case, except that the machinery taken from [26] has only been sketched.

7.8. Let us now give the remaining counter-examples. With almost no additional effort, we have

Counter-example 1: Let I be the β closed maximal ideal 7.5.4. Then I is neither principal, nor quasi-principal. Furthermore, the function

(1) $$z , \quad z \in R ,$$

is a function in $H^\infty(R)$ with no β interior-exterior factorization, even though the associated l.a.m.

(2) $$|z| , \quad z \in R ,$$

is outer.

We shall give a proof of counter-example 1 which is complete, except for those results from section 7.7 which were only sketched:

Proof modulo section 7.7: From counter-example 7.5.1, we know that I is not quasi-principal. To see that I is not principal either, assume the contrary. Let $I = \psi H^\infty(R)$, where $\psi \in H^\infty(R)$. Let I be the inner factor of ψ . Clearly

$$\psi H^\infty(R) \subseteq H^\infty(I:R) .$$

However, $z \in \psi H^\infty(R)$, and we know from lemma 7.5.1 that the inner factor of z is 1 . Thanks to lemma 7.2.1 and equation 7.2.1, $I = 1$. Thus $|\psi|$ is outer, and by theorem 7.7.2C, ψ is a unit in $H^\infty(R)$. We conclude that $\psi H^\infty(R) = H^\infty(R)$, which is a contradiction.

As for the remaining assertions of counter-example 1, let ψF be the interior-exterior factorization of the function defined by 1. Let I be the inner factor of $|\psi|$. Thanks to the β exteriority of F,

$$\mathrm{Cl}\, zH^\infty(R) \subseteq \psi H^\infty(R) = H^\infty(I:R) .$$

As before, this shows $I = 1$, so ψ is a unit in $H^\infty(R)$. Thus $z = \psi F$ is also β exterior, which is a contradiction.

The second counter-example we shall give will yield an example of a quasi-principal submodule in $H^p(R)$ which is not principal, even though R is admissible. Because of the length of the construction and the developments leading up to it, we shall only give a sketch.

<u>Lemma</u> 1: Let $u \in Q(R)$. Let Γ_α , $\alpha \in A$, be a univalent listing of the closed contours of Γ , and suppose $u^* = 0$ on $\Gamma \sim \bigcup\{\Gamma_\alpha : \alpha \in A\}$. For each $\alpha \in A$, let w_α be the bounded solution to the Dirichlet problem for the function equal to 1 on Γ_α and 0 on $\Gamma \sim \Gamma_\alpha$. Let $j \in A$, and let u^* be integrable with respect to the total variation measure $d\mu_j$ of $*dw_j$ on Γ . Then

$$\int_{\gamma_j} *du = \int_\Gamma u^* dw_j , \tag{3}$$

for each cycle γ_j in R homologous to Γ_j in $R \cup \Gamma$.

<u>Sketch of the proof</u>: If $R \cup \Gamma$ is compact, that is if R is conformally equivalent to a Jordan region, and if u may be continuously extended to $R \cup \Gamma$,

then equation 3 follows quickly from Green's theorem. The proof of lemma 1 for general R but for $u \in HB(R)$ may be reduced to this special case by considering exhaustions of R by Jordan subregions. The proof of lemma 1 for $u \in Q(R)$ then follows by considering increasing sequences of bounded harmonic functions.

Lemma 1 actually holds if $R \cup \Gamma$ is merely a harmonically complete bordered Riemann surface [26, p. 388]. However, in the next two lemmas, we assume that $R \cup \Gamma$ is a harmonically complete bordered simple region:

<u>Lemma</u> 2: Let $u \in HB(R)$. In the notation of lemma 1,

$$(4) \qquad \left|\int_{\gamma_j} *du\right| \le 2\|u\|_\infty \int_{\bigcup\{\Gamma_\alpha \,:\, \alpha \in A\}} d\mu_j \quad ,$$

for each $j \in A$.

Lemma 2 follows immediately from lemma 1. Suppose now that R is actually an admissible surface, as well as the interior of a harmonically complete bordered simple region. In addition, suppose R is infinitely connected and the exceptional set E of R has harmonic measure zero with respect to the Jordan subregion of $\hat{\mathbb{C}}$,

$$(5) \qquad \hat{\mathbb{C}} \sim \bigcup\{A_i : A_i \cap E = \emptyset\} .$$

Here we have used the notation of definition 7.7.1. Then we have

<u>Lemma</u> 3: In the notation of lemma 1, suppose there exists an infinite subset B of A such that

$$(6) \qquad \lim_{\substack{i \in B \\ i \to \infty}} \int_{\bigcup\{\Gamma_\alpha :\, \alpha \in A\}} d\mu_i = 0 .$$

Then there exists at least one bounded inner l.a.m. I such that each quasi-principal submodule $H^p(I:R)$ is not principal.

Sketch of the proof: In view of theorem 7.6.2, it suffices to construct a bounded inner l.a.m. I on R such that each $f \in H^\infty(R)$, for which $|f|$ has inner factor I, fails to be interior. To construct I, we first construct a sequence $b_i \in R$, $i = 1,2,\cdots$, satisfying the following two conditions: First,

(7) $$\sum_{i=1}^{\infty} G(b_i,z) < \infty ,$$

for each $z \neq$ any b_i. Second, there exists a univalent function $\sigma\colon \mathbb{Z}^+ \longrightarrow B$ such that for each integer $i > 0$,

(8) $$1/2 \leq \frac{-1}{2\pi} \int_{\Gamma_{\sigma(i)}} \sum_{j=1}^{n} {*dG}(b_j,z) \leq 3/4 ,$$

for each $n > i$. Let

$$H = \sum_{j=1}^{\infty} G(b_j,z) ,$$

and set

(9) $$I = \exp(-H) .$$

Suppose F is an outer l.a.m. such that IF has a global branch $f \in H^\infty(R)$. Suppose in addition that f is interior. Thanks to theorem 7.7.2C, there exist constants $0 < k < K < \infty$ such that $k < F < K$. Then $\ln F \in HB(R)$. But thanks to lemma 2 and equation 6,

(10) $$\lim_{i\to\infty} \left| \int_{\gamma_{\sigma(i)}} {*d} \ln F \right| = 0 .$$

On the other hand, $\theta_{IF} = \theta_I \theta_F = 1$. From equations 5.1.2 and 8, we conclude that the limit 10 cannot be zero. Thus f cannot be interior.

Lemma 4: In the notation of lemma 3, let f be any function in $H^\infty(R)$ such that $|f|$ has inner factor I. Then f fails to have either a β interior-exterior factorization, or a p interior-exterior factorization for any $p \in [1,\infty]$.

Lemma 4 actually guarantees the existence of a function $f \in H^\infty(R)$ such that f fails to have either a β interior-exterior factorization or a p interior-exterior factorization for any $p \in [1,\infty]$. To construct such a function, simply choose an appropriate l.a.m. ε from the system of multipliers 5.2.1, so that $I\varepsilon$ has a global branch f. Then f lies in $H^\infty(R)$ and $|f|$ has inner factor I.

Sketch of the proof of lemma 4: Clearly, an ∞ exterior function is also β exterior. Thus, thanks to theorem 7.6.1, if f has an ∞ interior-exterior factorization, it also has a β interior-exterior factorization. Suppose that $f = \psi F$ is either a β interior-exterior factorization or a p interior-exterior factorization for some $p \in [1,\infty]$. To fix ideas, assume the latter. Thanks to the exteriority of F,

$$\mathrm{Cl} fH^\infty(R) = \psi H^\infty(R) .$$

Clearly I is the g.c.i.d. of the inner factors of the l.a.m.'s in the set $\{|u|: u \in fH^\infty(R)\}$. Thanks to theorems 7.7.1 and 7.4.1, the smallest closed submodule containing $fH^\infty(R)$ is $H^p(I:R)$, so

$$\mathrm{Cl} fH^\infty(R) = H^p(I:R) .$$

Thus $H^p(I:R)$ is principal, which is impossible by lemma 3. The β interior-exterior case is handled in an entirely similar manner.

To complete the demonstration that theorem 7.1.1 cannot be improved to assert that each closed $H^\infty(R)$ submodule of $H^p(R)$ is principal for arbitrary admissible R, as well as the demonstration that interior-exterior factorizations may not exist for bounded analytic functions on an admissible surface, we must produce an example of a plane region satisfying the hypotheses of lemmas 3 and 4.

Counter-example 2: Let a_n, $n = 1,2,\cdots$, be a strictly increasing sequence on $[0,1)$ such that

(11) $$\sum_{i=1}^{\infty}(1 - a_i) < \infty .$$

Let r_n, $n = 1,2,\cdots$, be a sequence of strictly positive numbers such that $\mathrm{Cl}\Delta(a_n,r_n) \cap \mathrm{Cl}\Delta(a_m,r_m) = \emptyset$ if $n \neq m$, and $\mathrm{Cl}\Delta(a_n,r_n) \subseteq \Delta(0,1)$ for all n. Let s_n, $n = 1,2,\cdots$, be a sequence such that

(12) $$0 < s_n < r_n \exp(-2^n) ,$$

for each n. Let

(13) $$R = \Delta(0,1) \sim \bigcup \{\mathrm{Cl}\Delta(a_n,s_n): n = 1,2,\cdots\} .$$

Then R is an admissible surface, in fact a Blaschke region. Furthermore, R satisfies the remaining hypotheses of lemmas 3 and 4.

Sketch of the proof: We shall give only the briefest sketch. Clearly $\mathrm{Cl}R$ is a bordered simple region. The exceptional set E consists of only one point, the point 1, so $\mathrm{Cl}R = R \cup \Gamma$ is harmonically complete. Furthermore, E is a set of harmonic measure 0 with respect to $\Delta(0,1)$. The convergence of the series 11 guarantees that the a_i's are the zeros of a convergent Blaschke product, so R is a Blaschke region (c.f. definition 5.3.2 f.f.). Condition 6 is the one which is difficult to verify. To give the briefest sketch, we first let $w_{n,m}$ be the solution to the Dirichlet problem for the region

$$\hat{\mathbb{C}} \sim \mathrm{Cl}\Delta(a_m,s_m) \cap \mathrm{Cl}\Delta(a_n,s_n) ,$$

and the boundary data

$$w_{n,m} = 1 \quad \text{on} \quad \partial\Delta(a_n,s_n)$$
$$w_{n,m} = 0 \quad \text{on} \quad \partial\Delta(a_m,s_m) .$$

From the maximum principle, we obtain the estimate

$$w_{n,m}(z) \le (\ln r_m/s_m)^{-1}\ln(|z-a_m|/s_m) , \tag{14}$$

for $z \in \Delta(a_m,r_m) \sim \Delta(a_m,s_m)$. Since both sides of 14 vanish on $\partial\Delta(a_m,s_m)$, we see that on $\partial\Delta(a_m,s_m)$,

$$-2^{-m}d\theta \le *dw_{n,m} \le 0 , \tag{15}$$

by applying 12. We then use the maximum principle to obtain

$$0 \le w_n \le w_{n,m} , \tag{16}$$

on $R \cup \Gamma_{/}$ and thus on $\partial\Delta(a_m,s_m)$. Both sides of 16 also vanish on $\partial\Delta(a_m,s_m)$, and the measures $*dw_n$ and $*dw_{n,m}$ are both negative on $\partial\Delta(a_m,s_m)$. We may apply 15 and 16 together to obtain 6.

Chapter 8

Construction of Admissible Surfaces

8.1. We return to the theorems left unproved in section 5.3. Their domain of applicability may be extended somewhat by the following observations: A simple argument involving the Heine-Borel theorem and definition 5.3.2C shows that if (R_0,R) satisfies all the requirements of a Blaschke pair, except that R may not be connected, then R is automatically connected (c.f. [26, pp. 135-137 for a proof in the case where R_0 is a Jordan region in $\mathbb{C}$). If R_0 is a plane region, it is possible to show by fairly elementary means that condition C may be replaced by the simpler condition that $\mathbb{C} \sim A$ is always connected [28, theorem 9.1.2]. I am indebted to Robert Magedson for pointing out to me that if

$$R = \hat{\mathbb{C}} \sim \bigcup\{A_i:\ i = 1,2,\cdots\} ,$$

where A_i, $i = 1,2,\cdots$, is a sequence of disjoint compact continua such that $\hat{\mathbb{C}} \sim A_i$ is connected for each i, and $\hat{\mathbb{C}}$ is the extended complex plane, then R is connected if it is open.

Let us take up the unfinished business of proving theorems 5.3.2 and 5.3.3. Throughout the remainder of this chapter (R_0,R) shall be a Blaschke pair. For simplicity, all closures, interiors, and boundaries of subsets of R will be taken with respect to R_0 unless otherwise noted. We shall denote the Green's function for R_0 by $G_0(a,z)$ as usual, and R^α (and by extension R_0^α) shall have the meaning assigned by equation 6.2.7. Although we shall use parts of chapter 6, we shall only use those parts in which R_0 was merely assumed to be regular. By contrast with previous chapters, we shall abandon the convention that $1 \le p \le \infty$, $1 \le q \le \infty$, and $1/p + 1/q = 1$.

The remainder of this section will be devoted to a series of preparatory lemmas. Unless otherwise noted, the coefficient ring $\mathbb{K}$ for homology will be $\mathbb{R}$ throughout this section (c.f. the beginning of chapter 5). First, we shall need several connectivity arguments which reduce to the following trivial lemma:

<u>Lemma</u> 1: Let X be a Hausdorff space, and let V be a locally connected subset of X. Let $A \subseteq X$ and let B be a component of $A \cap V$ such that $\mathrm{Cl}_X B \subseteq V$. Then $\partial_X B \subseteq \partial_X A$. If A is locally connected, then B is a component of A and is open in A.

We shall omit the proof of lemma 1. Primarily, the arguments in which lemma 1 will be used involve the uniqueness of the solution to the Dirichlet problem on regions with compact closure, together with the minimum principle for superharmonic functions, as in

<u>Lemma</u> 2: Let A be a component of $R_0 \sim R$. Let V be an open subset of R_0 with compact closure such that $A \subseteq V$ and $(\mathrm{Cl}V) \cap (R_0 \sim R) = A$. Suppose $0 < \alpha < \inf\{G(a_0,z): z \in \partial V\}$. Then $(R_0 \sim R^\alpha) \cap V$ is connected.

<u>Proof</u>: Suppose B is a component of $(R_0 \sim R^\alpha) \cap V$ not containing A. Since $(R_0 \sim R^\alpha) \cap V$ is open, B is also open. Clearly $\mathrm{Cl}B \subseteq V$. Since $R_0 \sim R^\alpha$ is locally connected, B is a component of $R_0 \sim R^\alpha$ by lemma 1. In particular, B is closed in $R_0 \sim R^\alpha$, so $\mathrm{Cl}B \cap A = \phi$. Thus $\mathrm{Cl}B$ is a compact subset of R. Thanks to lemma 1 again, $\partial B \subseteq \partial(R_0 \sim R^\alpha)$, that is $G(a_0,z) = \alpha$ on ∂B. Hence $G(a_0,z) \equiv \alpha$ on B. By the minimum principle, $G(a_0,z) \equiv \alpha$ on R, which is impossible.

As canonical notation, let

$\beta(S)$ = the first Betti number of S

$n(S)$ = the zeroth Betti number of S

$\chi(S)$ = the Euler characteristic of S ,

for each topological space S on which all Betti numbers are finite and all but finitely many Betti numbers vanish. The symbol β standing alone will never represent a Betti number, so no confusion can arise between $\beta(S)$ and the β topology on S. A corollary to Hopf's classical theorem on the singularities of vector fields will play a crucial role in the sequel:

Lemma 3: Let S be a compact bordered Riemann surface. Let h be harmonic on S, equal to a constant α_0 on some of the contours of the border of S, and equal to another constant α_1 on the remaining contours. Then the number of critical points of h, counting multiplicity, is $-\chi(S)$

Elegant elementary proofs of lemma 3 also exist [31, pp. 32-36]. The first applications of lemma 3 will be to the production of conformal annuli. Let $0 < \alpha < \beta$ and let V be a component of $\text{Int}(R^\alpha \sim R^\beta)$ containing no critical points of $G(a_0,z)$. Let $\alpha < \gamma < \delta < \beta$, and let W be a component of $\text{Cl}(R^\gamma \sim R^\delta) \cap V$. From lemma 1, we obtain

$$\partial W \subseteq \partial \text{Cl}(R^\gamma \sim R^\delta) .$$

Since V contains no critical points, there exists a covering of ∂W by parametric disks similar to the covering 6.2.9. The set W is also closed, and hence compact, so W is a closed Jordan subregion of R. As in the proof of lemma 2, we have

$$\begin{aligned} &\partial W \cap \{z: \ G(a_0,z) = \gamma\} \neq \phi \\ (1) \qquad & \\ &\partial W \cap \{z: \ G(a_0,z) = \delta\} \neq \phi . \end{aligned}$$

Thanks to lemma 3, $\chi(W) = 0$. By the classification theorem for surfaces, W is homeomorphic to a closed annulus [1, pp. 90-104]. We have almost proved

Lemma 4: Let V be as above. Then V is conformally equivalent to an annulus.

Proof: Note that ∂V may contain critical points of $G(a_0,z)$. However, for $\alpha < \gamma < \delta < \beta$, we have just shown that each component of

$$U^{\gamma\delta} = \mathrm{Cl}(R^{\gamma} \sim R^{\delta}) \cap V ,$$

is a closed Jordan subregion of R which is homeomorphic to a closed annulus. If $U^{\gamma\delta}$ were connected, then

$$H_1(U^{\gamma\delta}) = \mathbb{R} .$$

The proof will reduce to showing that $U^{\gamma\delta}$ is connected for all γ and δ sufficiently close to α and β respectively.

First let $\alpha < \gamma < \varepsilon < \zeta < \delta < \beta$. Thanks to equation 1 and the intermediate value theorem, each component of $U^{\gamma\delta}$ intersects $U^{\varepsilon\zeta}$. Thus

(2) $$n(U^{\gamma\delta}) \leq n(U^{\varepsilon\zeta}) .$$

However, V is connected, so

(3) $$\mathbb{R} = H_0(V) = \varinjlim H_0(U^{\gamma\delta}) ,$$

where the omitted maps are induced by inclusion maps, and the set of pairs (γ,δ) is directed in the obvious way. Combining 2 and 3, we see that there exists a pair (γ_0,δ_0) such that $\alpha < \gamma_0 < \delta_0 < \beta$ and

(4) $$H_0(U^{\gamma\delta}) = \mathbb{R} ,$$

for all $\alpha < \gamma < \gamma_0$ and $\delta_0 < \delta < \beta$. For such pairs (γ,δ), $U^{\gamma\delta}$ is connected. Thus

$$\mathbb{R} = \varinjlim H_1(U^{\gamma\delta}) = H_1(V) .$$

Since V is orientable, $H_1(V, \mathbb{Z})$ has no torsion, so

(5) $$H_1(V, \mathbb{Z}) = \mathbb{Z} .$$

Thanks to equation 5, the differential $*dcG(a_0,z)$ has period 0 or $\pm 2\pi$ on V for some constant $c < 0$. Applying equation 5.1.2, we see that the l.a.m. $\exp(cG(a_0,z))$ has a branch f on V. The map

(6) $$\psi: V \longrightarrow \Delta(0,1) \sim \mathrm{Cl}\Delta(0,r) ,$$

defined by $\psi(z) = sf(z)$, $z \in V$, for a suitable choice of the constant s, is a conformal homeomorphism.

We still must show that the map ψ is $1-1$ and onto. For $\alpha < \varepsilon < \beta$, the set $\partial R^{\varepsilon} \cap V$ consists of exactly one contour. To see why, let $\alpha < \gamma < \gamma_0$, $\delta_0 < \delta < \beta$, and let $\gamma < \varepsilon < \delta$. Since $U^{\gamma\delta}$ is homeomorphic to a closed annulus, $\partial R^{\gamma} \cap V$ and $\partial R^{\delta} \cap V$ each consist of only one contour. Here, we have used the fact that a homeomorphism between bordered surfaces carries the border of one onto the border of the other [1, p. 23]. An application of equation 1 to each component of $U^{\gamma\varepsilon}$ shows that $U^{\gamma\varepsilon}$ is connected, so $\partial R^{\varepsilon} \cap V$ also consists of only one contour. A simple argument involving the positivity of the measure $*dG(a_0,z)$ on the contour $\partial R^{\varepsilon} \cap V$ (with the proper orientation) shows that ψ is $1-1$ and onto.

In the course of the proof of lemma 4, we have obtained the following useful result: For each pair (γ,δ), where $\alpha < \gamma < \delta < \beta$,

(7) $$\psi(U^{\gamma\delta}) = \mathrm{Cl}\Delta(0,r_1) \sim \Delta(0,r_2) ,$$

for some pair (r_1,r_2), where $r < r_2 < r_1 < 1$. In addition,

(8) $$\psi(\partial R^{\gamma} \cap V) = \partial\Delta(0,r_1) ,$$
$$\psi(\partial R^{\delta} \cap V) = \partial\Delta(0,r_2) .$$

Here, the notation is as in the proof of lemma 4.

Equations 7 and 8 may be applied to show that inclusion maps are actually excisions. For example, let $\delta > 0$ be a non-critical value of $G(a_0,z)$. Let F be a component of $Cl(R_0 \sim R^{\delta})$, and let W be a subset of R_0 containing F . Then the injection of space pairs

$$(9) \qquad p: \ (W \sim \operatorname{Int} F, \ \partial F) \longrightarrow (W,F) \ ,$$

is an excision. To see why, recall from the proof of lemma 6.2.2 that the critical values of $G(a_0,z)$ form a discrete subset of $\mathbb{R}$. Choose α and β so that $0 < \alpha < \delta < \beta$, and no critical values of $G(a_0,z)$ occur in the interval $[\ \alpha,\beta\ ]$. Let $\alpha < \gamma < \delta$. Let F^{γ} be the union of the components of $Cl(R_0 \sim R^{\gamma})$ contained in F . Thanks to the covering 6.2.9, each component of $Cl(F \sim F^{\gamma})$ is a component of $Cl(R^{\gamma} \sim R^{\delta})$, and is open as well as closed in $Cl(R^{\gamma} \sim R^{\delta})$. By compactness, there are only finitely many components of $Cl(F \sim F^{\gamma})$. Each of these is contained in a component of $\operatorname{Int}(R^{\alpha} \sim R^{\beta})$. Applying equations 7 and 8 to each of the components of $Cl(F \sim F^{\gamma})$, we see that F^{γ} is a strong deformation retract of F . However, the injection

$$q: \ (W \sim F^{\gamma}, \ F \sim F^{\gamma}) \longrightarrow (W,F)$$

is an excision. Hence p is also an excision. We are now in a position to prove the following lemma. Here $\mathcal{A}$ is the set of components of $R_0 \sim R$:

Lemma 5: There is a family

$$B_A \ , \ A \in \mathcal{A} \ ,$$

of open subsets of R with disjoint compact closures in R_0 , such that B_A is conformally equivalent to an annulus, and such that $A \cup B_A$ is a neighborhood of A , for each component $A \in \mathcal{A}$.

<u>Proof</u>: Thanks to definition 5.3.2B, it suffices to prove the following: For each component $A \in \mathcal{A}$, and for each connected open neighborhood V of A with compact closure such that $\mathrm{Cl}V \cap (R_0 \sim R) = A$, there exists an open subset U of V containing A such that $U \sim A$ is conformally equivalent to an annulus. Let V be such an open neighborhood. By lemma 2, for

$$0 < \alpha < \inf\{G(a_0,z):\ z \in \partial V\} ,$$

the only component of $(R_0 \sim R^\alpha) \cap V$ is the component of $R_0 \sim R^\alpha$ containing A . Call this component W_α . We shall set $U = W_\alpha$ for some sufficiently small α . As in the case of lemma 4, the proof reduces to showing $H_1(W_\alpha \sim A) = \mathbb{R}$ for α sufficiently small.

We shall only consider W_α for α not a critical value of $G(a_0,z)$. Thanks to the parametric disks 6.2.9, each W_α is a Jordan region of R_0 . Similarly, each of the finitely many components of $\mathrm{Cl}W_\alpha \sim W_\beta$ is a closed Jordan region of R_0 for $0 < \beta < \alpha$. Consider the space pair $(\mathrm{Cl}W_\alpha,\mathrm{Cl}W_\beta)$, where $0 < \beta < \alpha$. Since the map 9 is an excision,

$$H_2(\mathrm{Cl}W_\alpha,\mathrm{Cl}W_\beta) = H_2(\mathrm{Cl}W_\alpha \sim W_\beta,\partial W_\beta) .$$

As in the derivation of equation 1, the border of each component of $\mathrm{Cl}W_\alpha \sim W_\beta$ intersects both ∂W_α and ∂W_β . Thus

$$H_2(\mathrm{Cl}W_\alpha \sim W_\beta,\partial W_\beta) = 0 . \tag{10}$$

The proof of 10 is almost identical to the proof that H_2 of a bordered Riemann surface vanishes. (C.f. [1, p. 53] for an elementary proof involving simplicial homology. A similar argument occurs in [48]). From the homology exact sequence for the pair $(\mathrm{Cl}W_\alpha,\mathrm{Cl}W_\beta)$, we obtain the exact sequence

$$0 \longrightarrow H_1(\mathrm{Cl}W_\beta) \longrightarrow H_1(\mathrm{Cl}W_\alpha) .$$

Consequently $\beta(\mathrm{Cl}W_\alpha)$ is an increasing non-negative integer valued function of α . Hence for some non-critical $\alpha_0 > 0$, $\beta(\mathrm{Cl}W_\alpha)$ is constant for all non critical α such that $0 < \alpha < \alpha_0$.

Fix $\alpha \in (0,\alpha_0)$, and suppose α is not a critical value of $G(a_0,z)$. Recall from definition 5.3.2C that the connectedness of $\mathrm{Cl}W_\alpha$ implies the connectedness of $\mathrm{Cl}W_\alpha \sim A$. As in the derivation of equation 10, $n(\mathrm{Cl}W_\alpha \sim W_\beta)$ is an increasing function of β . As in the derivation of equation 4, there exists $0 < \beta_0 < \alpha$ such that $\mathrm{Cl}(W_\alpha \sim W_\beta)$ is connected for $0 < \beta < \beta_0$. Thanks to equations 7 and 8, $(\mathrm{Cl}W_\alpha, \mathrm{Cl}W_\beta, \mathrm{Cl}W_\alpha \sim W_\beta)$ is an exact triad for $0 < \beta < \alpha$. An application of the Mayer-Vietoris sequence shows that

$$\text{(11)} \qquad \chi(\mathrm{Cl}W_\alpha) = \chi(\mathrm{Cl}W_\beta) + \chi(\mathrm{Cl}W_\alpha \sim W_\beta)$$

Here we have used

$$\chi(\mathrm{Cl}W_\beta \cap \mathrm{Cl}W_\alpha \sim W_\beta) = \chi(\partial W_\beta) = 0 \quad .$$

Because the function $t \longrightarrow \chi(\mathrm{Cl}W_t)$ is constant for $0 < t < \alpha_0$, we have

$$\chi(\mathrm{Cl}W_\alpha \sim W_\beta) = 0 ,$$

by equation 11. Consequently $H_1(\mathrm{Cl}W_\alpha \sim W_\beta) = \mathbb{R}$ for $0 < \beta < \beta_0$. Passing to a direct limit, we obtain

$$\text{(12)} \qquad \mathbb{R} = \varinjlim H_1(\mathrm{Cl}W_\alpha \sim W_\beta) = H_1(\mathrm{Cl}W_\alpha \sim A) ,$$

where the omitted maps are induced by inclusion maps and the set of non-critical β in $(0,\beta_0)$ is ordered in the obvious way. A similar passage to a direct limit shows that $H_1(W_{\alpha_0} \sim A) = \mathbb{R}$. As in the construction of the map 6, we may employ $G(a_0,z)$ to construct a conformal homeomorphism

(13) $$\psi:\ W_{\alpha_0} \sim A \xrightarrow{\text{onto}} \Delta(0,1) \sim \mathrm{Cl}\Delta(0,r) ,$$

for some $0 < r < 1$. Choose a non-critical α such that $0 < \alpha < \alpha_0$, and set $U = W_\alpha$. Then ψ maps $U \sim A$ conformally and univalently onto $\Delta(0,s) \sim \mathrm{Cl}\Delta(0,r)$ for some $r < s < 1$. This completes the proof of lemma 5.

For each component A of $R_0 \sim R$, we have explicitly constructed $B_A = U \sim A$, so the map 13, together with the analogues of equations 7 and 8, may be used to construct an annular collar around the contour $\partial_R B_A$. As in the proof that the map 9 is an excision, we immediately have

Lemma 6: Let

(14) $$B_A\ ,\quad A \in \mathcal{A}\ ,$$

be the family constructed in the proof of lemma 5. Let

$$C = \{B_A:\ A \in \mathcal{A}\}$$

$$D = \{B_A \cup A:\ A \in \mathcal{A}\} .$$

Then $R \sim C = R_0 \sim D$, when endowed with the conformal structure inherited from R_0 , is a bordered Riemann surface with border $\partial_R C = \partial D$. Furthermore, the injections of space pairs

(15) $$p:\ (R \sim C, \partial_R C) \longrightarrow (R,C) ,$$

$$q:\ (R_0 \sim D, \partial D) \longrightarrow (R,D) ,$$

are both excisions.

8.2. We interrupt the chain of topological considerations to give three lemmas of a pleasantly analytic nature. In the first of these, $\mathcal{A}$ shall once again be the set of components of $R_0 \sim R$.

Lemma 1: There exists a family of positive Baire probability measures on R_0,

(1) $$\mu_A \ , \quad A \in \mathcal{A} \ ,$$

such that μ_A has support contained in A for each $A \in \mathcal{A}$. Furthermore, the function

(2) $$h_A(z) = \int G_0(a,z)d\mu_A(a) \ , \quad z \in R \ ,$$

lies in $h^\infty(R)$ for each $A \in \mathcal{A}$, and the series

(3) $$\sum(h_A(z): \ A \in \mathcal{A}) \quad < \infty \ ,$$

for each $z \in R$.

Proof: Let A_i, $i = 1,2,\cdots$, be a univalent enumeration of the components of $R_0 \sim R$, and let $a_i = a_{A_i}$, $i = 1,2,\cdots$, be as in definition 5.3.2A. For fixed i, let V be a neighborhood of a_i such that

$$|G(a_0,a) - G(a_0,a_i)| < 2^{-i} \ ,$$

for all $a \in V$. Since A_i is a continuum, there is a continuum B contained in $V \cap A_i$. Let h be the normal solution to the Dirichlet problem for $R_0 \sim B$ with boundary conditions $h(p) = 1$, $p \in \partial B$. (The normal solution also satisfies $h \geq 0$ and $\lim_{z\to\infty} h(z) = 0$, c.f. [5, p. 21]). Set $h = 1$ on B. Then h is continuous and superharmonic on R_0. Clearly L. H. M. $h = 1$. By the Riesz representation theorem (c.f. [5, p. 41]),

$$h(z) = \int G_0(a,z)d\mu(a) \ , \quad z \in R_0 \sim B \ ,$$

where μ is a positive Baire measure on R_0. Since h is harmonic on $R_0 \sim B$, μ has support contained in B. Of course, $\mu(B) > 0$. Set

$$\mu_{A_i} = \mu_i = \mu/\mu(A_i) \quad .$$

Clearly μ_i is a probability measure. Thanks to equation 2, we have

$$(4) \qquad h_{A_i}(z) = \frac{1}{\mu(A_i)} h(z) \ , \quad z \in R \ ,$$

so the function h_{A_i} is harmonic and positive on R, and bounded above by $1/\mu(A_i)$. Also,

$$\sum_{i=1}^{\infty}\int G_0(a,a_0)d\mu_i(a) \le \sum_{i=1}^{\infty}\int (G_0(a_0,a_i)+2^{-i})d\mu_i(a)$$
$$= \sum_{i=1}^{\infty}(G_0(a_0,a_i)+2^{-1}) < \infty \quad ,$$

thanks to definition 5.3.2A. By the Harnack monotone convergence theorem, inequality 3 holds for each $z \in R$.

Lemma 2: Let S be arbitrary Riemann surface. Let f_n , $n = 1,2,\cdots$ be a sequence of l.a.m.'s on S which is bounded on each compact subset of S . Then a subsequence $f_{\sigma(n)}$, $n = 1,2,\cdots$, converges uniformly on compact subsets of S to an l.a.m. f .

Proof: By the Heine-Borel theorem and a standard, classical normal families argument (c.f. [20, p. 246]), it suffices to prove the following: Let (V,ψ) be a parametric disk in S . Then there exists a subsequence $f_{\sigma(n)}$, $n = 1,2,\cdots$, converging uniformly on V to an l.a.m. f . This, however, is trivial. Let (W,ψ) be a parametric disk such that W contains $\text{Cl}V$. On W , f_n has a branch g_n for each n . Exploiting the normality of the sequence g_n , $n = 1,2,\cdots$, we obtain the existence of the required subsequence $f_{\sigma(n)}$, $n = 1,2,\cdots$.

From lemma 2, we easily obtain.

Lemma 3: Let f_n , $n = 1,2,\cdots$, be a sequence of l.a.m.'s converging pointwise boundedly on compact subsets of S to a function f . Then f is an l.a.m. and $\theta_{f_n} \to \theta_f$ pointwise on $H_1(S, \mathbb{Z})$.

<u>Proof</u>: Each subsequence $f_{\sigma(n)}$, $n = 1,2,\cdots$, contains a subsequence $f_{\tau\circ\sigma(n)}$, $n = 1,2,\cdots$, converging uniformly on compact subsets of S to an l.a.m. g . From the analytic continuation process used to construct the character of an l.a.m. in section 5.1, and from the proof of lemma 2, we see that given $\gamma \in H_1(S, \mathbb{Z})$, the sequence $f_{\tau\circ\sigma(n)}$, $n = 1,2,\cdots$, can be chosen so

$$\lim_{n\to\infty} \theta_{f_{\tau\circ\sigma(n)}}(\gamma) = \theta_g(\gamma) .$$

The usual induced convergence argument (c.f. [20, p. 251]) shows that $g = f$ and $\theta_{f_n} \to \theta_f$ pointwise on $H_1(S)$.

8.3. We are now ready to begin the proof of theorem 5.3.2. Throughout this section, $\mathcal{A}$ shall be the set

$$\{B_A \cup A\colon\ A \text{ is a component of } R_0 \sim R\} ,$$

and $\mathcal{B}$ shall be the set

$$\{B_A\colon\ A \text{ is a component of } R_0 \sim R\} .$$

Here the family of B_A's is the family of conformal annuli 8.1.14. Construct an auxiliary Riemann surface $R_1 \supseteq R$ by welding a closed conformal disk C_A to each annulus B_A . Let D_A be the open conformal disk $C_A \cup B_A$, and let $\mathcal{D}$ be the set

$$\{D_A\colon\ A \text{ is a component of } R_0 \sim R\} .$$

By contrast with previous sections, we shall let the coefficient ring $\mathbb{K}$ for homology be $\mathbb{Z}$.

Proof of Theorem 5.3.2: First we shall decompose $H_1(R)$ into the direct sum of $H_1(R_1)$ and $H_1(\cup\mathcal{B})$. Consider the commutative diagram of inclusion maps

(1)
$$\begin{array}{ccc} \cup\mathcal{D} & \xrightarrow{\ i_{\mathcal{D}}\ } & R_1 \\ \uparrow{\scriptstyle i_3} & & \uparrow{\scriptstyle i_1} \\ \cup\mathcal{B} & \xrightarrow{\ i_{\mathcal{B}}\ } & R \\ \downarrow{\scriptstyle i_2} & & \downarrow{\scriptstyle i_0} \\ \cup A & \xrightarrow{\ i_A\ } & R_0 \end{array}$$

Diagram 1 induces the commutative diagram

(2)
$$\begin{array}{ccccccc} H_1(\cup\mathcal{D}) & \xrightarrow{\ i_{\mathcal{D}}\ } & H_1(R_1) & \xrightarrow{\ j_{\mathcal{D}}\ } & H_1(R_1,\cup\mathcal{D}) & \xrightarrow{\ \partial_{\mathcal{D}}\ } & H_0(\cup\mathcal{D}) \\ \uparrow{\scriptstyle i_3} & & \uparrow{\scriptstyle i_1} & & \uparrow{\scriptstyle k_1} & & \uparrow{\scriptstyle i_3} \\ H_1(\cup\mathcal{B}) & \xrightarrow{\ i_{\mathcal{B}}\ } & H_1(R) & \xrightarrow{\ j_{\mathcal{B}}\ } & H_1(R,\cup\mathcal{B}) & \xrightarrow{\ \partial_{\mathcal{B}}\ } & H_0(\cup\mathcal{B}) \\ \downarrow{\scriptstyle i_2} & & \downarrow{\scriptstyle i_0} & & \downarrow{\scriptstyle k_0} & & \downarrow{\scriptstyle i_2} \\ H_1(\cup A) & \xrightarrow{\ A\ } & H_1(R_0) & \xrightarrow{\ j_A\ } & H_1(R_0,\cup A) & \xrightarrow{\ \partial_A\ } & H_0(\cup A) \end{array} .$$

The rows of diagram 2 are the homology exact sequences for the space pairs $(R_1,\cup\mathcal{D})$, $(R,\cup\mathcal{B})$, and $(R_0,\cup A)$ respectively. Here i_1 denotes the map induced in homology by the inclusion map i_1 in diagram 1, etc. The isomorphisms k_0 and k_1 are induced by the excisions (i_0,i_2) and (i_1,i_3) respectively. Also i_2 and i_3 are isomorphisms in dimension 0 because the inclusion maps i_2 and i_3 in diagram 1 carry different components of $\cup\mathcal{B}$ into different components of $\cup A$ and $\cup\mathcal{D}$ respectively.

Set $G_A = j_A(H_1(R_0))$, $G_{\mathcal{B}} = j_{\mathcal{B}}(H_1(R))$, and $G_{\mathcal{D}} = j_{\mathcal{D}}(H_1(R_1))$. By exactness, $k_1(G_{\mathcal{B}}) = G_{\mathcal{D}}$ and $k_0(G_{\mathcal{B}}) = G_A$. But $H_1(\cup\mathcal{D}) = 0$, so $G_{\mathcal{D}}$, and thus $G_{\mathcal{B}}$, are isomorphic to the free group $H_1(R_1)$ [1, p. 64]. In addition,

(3) $$H_2(R_0,\cup A) = H_2(R\ ,\cup B) = 0$$

Suppose 3 has been established. From the homology exact sequence, we obtain the commutative diagram of split exact sequences

(4) $$\begin{array}{ccccccccc}
0 & \longrightarrow & H_1(\cup B) & \xrightarrow{i_B} & H_1(R) & \underset{p_B}{\overset{j_B}{\rightleftarrows}} & G_B & \longrightarrow & 0 \\
 & & \downarrow{\scriptstyle i_2} & & \downarrow{\scriptstyle i_0} & & \downarrow{\scriptstyle k} & & \\
0 & \longrightarrow & H_1(\cup A) & \xrightarrow{i_A} & H_1(R_0) & \underset{p_A}{\overset{j_A}{\rightleftarrows}} & G_A & \longrightarrow & 0
\end{array} \; .$$

The existence of the splitting homomorphism p_B is guaranteed by the projectivity of G_B , and the splitting homomorphism p_A is defined to be $i_0 \circ p_B \circ k^{-1}$. Here k is the restriction of k_0 to G_B .

We still must show equation 3 is true. Thanks to lemma 8.1.6, $H_2(R_0,\cup A)$ and $H_2(R,\cup B)$ are both equal to $H_2(R \sim \cup B, \partial_R(\cup B))$. However, $R \sim \cup B$ is a non-compact bordered Riemann surface. As in the derivation of equation 8.1.10, the standard proof that H_2 of an open Riemann surface vanishes may be adapted to show that $H_2(R \sim \cup B, \partial_A \cup B)$ vanishes. This completes the construction of diagram 4.

We now interrupt the proof of theorem 5.3.2 for two lemmas. Throughout this section, if S is a topological space, $\chi(S)$ shall denote the group of characters on $H_1(S)$. We shall also assume that R_0 is admissible and that

(5) $$\varepsilon_\chi \, , \quad \chi \in \chi(R_0) \, ,$$

is a system of multipliers admitted by R_0 .

<u>Lemma</u> 1: There exists a family of outer l.a.m.'s,

$$f_\chi \, , \quad \chi \in \chi(R)$$

such that $0 < f_\chi \leq 1$ and

$$\theta_{f_\chi} \circ i_B = \chi \circ i_B ,$$

for each $\chi \in \mathcal{X}(R)$. Further, $f_1 = 1$.

Proof: Thanks to direct decomposition, $H_1(\cup B)$ is the direct sum of the family

$$(H_1(B_A), i_A), \quad A \text{ is a component of } R_0 \sim R .$$

Here i_A is the map induced by the obvious inclusion map. Appealing to lemma 8.1.5, we see that each $H_1(B_A)$ is cyclic, and is generated by the homology class of a cycle of the form

$$Z_A(t) = \psi(ue^{-2\pi it}) , \quad 0 \leq t \leq 2\pi ,$$

where ψ is the map 8.1.13 and $r < u < s$. For simplicity, let $\chi(Z_A) = \chi \circ i_B \circ i_A([Z_A])$ for each $\chi \in \mathcal{X}(R)$. Here $[Z_A]$ temporarily denotes the homology class of Z_A. Set $V = \psi^{-1}(\mathrm{Cl}\Delta(0,u) \sim \Delta(0,r)) \cup A$. Then V is a compact bordered Riemann surface with border consisting of one contour, the image of Z_A. Further, Z_A is oriented positively with respect to the interior of V. Thanks to equation 3.1.1,

$$(6) \qquad \int_{Z_A} *dG_0(a,z) = \begin{cases} -2\pi , & a \in A \\ 0 , & a \notin V \end{cases} .$$

Fix $\chi \in \mathcal{X}(R)$. For each component A of $R_0 \sim R$, let h_A and μ_A be as in lemma 8.2.1. Applying lemma 6.3.2 and equation 6, we obtain

$$(7) \qquad \int_{Z_B} *dh_A = \int_A \left(\int_{Z_B} *dG(a,z)\right) d\mu_A(a) = \begin{cases} -2\pi , & A = B \\ 0 , & A \neq B \end{cases} ,$$

for each pair of components A and B. Here we have used the fact that μ_A has unit mass and support contained in A. For each A, set

(8) $$\lambda_A = \frac{1}{2\pi} \arg \chi(Z_A) ,$$

where $0 \leq \lambda_A \leq 1$. Consider the function

(9) $$h = \sum \lambda_A h_A ,$$

where the sum is taken over all components A of $R_0 \sim R$. By the Harnack monotone convergence theorem and inequality 8.2.3, h is harmonic on R. Because the series 9 converges uniformly on compact subsets of R,

(10) $$\int_{Z_A} *dh = -2\pi\lambda_A ,$$

for each A, thanks to equation 7. Set

(11) $$f_\chi = \exp(-h) .$$

Thanks to equation 5.1.2 and equation 10, we have

$$\theta_{f_\chi} \circ i_B = \chi \circ i_B .$$

Clearly $f_1 = 1$. It remains to show that f_χ is outer for each $\chi \in \mathcal{X}(R)$. Thanks to lemma 8.2.1, each h_A is bounded and positive. From equation 9, we see that h is positive and quasi-bounded. Therefore, f_χ is outer and $0 < f_\chi \leq 1$.

Unfortunately, we have no control over the character of f_χ on $p_B(G_B)$. To correct for this, we must use the system of multipliers admitted by R_0 together with certain naturality properties of the character of an l.a.m.. If S_0 and S_1

are Riemann surfaces and $\alpha: S_0 \longrightarrow S_1$ is analytic, then

$$\theta_{u o \alpha} = \theta_u \circ \alpha_* \tag{12}$$

for each l.a.m. u on S_1. Here α_* is the map induced in homology by α. We shall only need equation 12 for the case where α is an inclusion map. In this case, equation 12 follows trivially from the analytic continuation process used to construct the character of an l.a.m. in section 5.1.

Define a function $\tau: \mathcal{X}(R) \longrightarrow \mathcal{X}(R_0)$ by

$$\tau(\chi) \circ p_A = (\chi o p_B o k^{-1})(\theta_{f_\chi} o p_B o k^{-1})^{-1} \tag{13}$$

$$\tau(\chi) \circ i_A = 1$$

Such a character exists on $H_1(R_0)$ thanks to the direct sum decomposition given by diagram 4. Consider the family of l.a.m.'s,

$$\delta_\chi = \varepsilon_{\tau(\chi)} f_\chi \ , \quad \chi \in \mathcal{X}(R) \ . \tag{14}$$

Lemma 2: For each $\chi \in \mathcal{X}(R)$, δ_χ is outer, $0 < \delta_\chi \le 1$, and $\theta_{\delta_\chi} = \chi$.

Proof: That δ_χ is outer and $0 < \delta_\chi \le 1$ follows immediately from equation 13 and the corresponding facts about f_χ and $\varepsilon_{\tau(\chi)}$. For convenience, set $f = f_\chi$, $\delta = \delta_\chi$, $\varepsilon = \varepsilon_{\tau(\chi)}$, and $\tau = \tau(\chi)$. From diagrams 1 and 4 and equations 12 and 13, we obtain

$$\theta_{\varepsilon o i_0} \circ i_B = \tau \circ i_A \circ i_2 = 1, \quad \theta_{\varepsilon o i_0} \circ p_B = \tau \circ p_A \circ k = (\chi o p_B)(\theta_f o p_B)^{-1} \ . \tag{15}$$

Thus

$$\theta_{\varepsilon f} \circ i_B = (\theta_{\varepsilon o i_0} o i_B)(\theta_f o i_B) = \chi \circ i_B \ ,$$

$$\theta_{\varepsilon f} \circ p_B = (\chi o p_B)(\theta_f o p_B)^{-1}(\theta_f o p_B) = \chi \circ p_B \ .$$

Since diagram 4 yields a direct sum decomposition of $H_1(R)$, $\theta_{\varepsilon f} = \chi$ on all of $H_1(R)$.

Lemma 2 shows that the family 14 satisfies parts A and B of definition 5.2.1. In fact, the family 14 satisfies the remaining part of definition 5.2.1, and so is a system of multipliers for R . The proof of this will conclude the proof of theorem 5.3.2.

Proof of Theorem 5.3.2 continued: Consider p sequences in $\mathcal{X}(R)$,

$$\chi(j,n): \ n = 1,2,\cdots , \quad j = 1,2,\cdots p ,$$

such that the sequence

$$\delta_{\chi(1,n)} \cdots \delta_{\chi(p,n)} , \ n = 1,2,\cdots ,$$

converges to an l.a.m. with global branch F . We must show that F is exterior.

For convenience, set $\varepsilon_{\tau(\chi(j,n))} = g_{j,n}$, $f_{\chi(j,n)} = f_{j,n}$, and $\tau(\chi(j,n)) = \tau(j,n)$ in equations 5, 11, and 13 respectively. Set $\lambda_A = \lambda_A^{j,n}$ and $h = h_{j,n}$ in equations 8 and 9 for $\chi = \chi(j,n)$. Let $A(i)$, $i = 1,2,\cdots$ be a univalent enumeration of the components of $R_0 \sim R$.

Thanks to lemma 8.2.2, by taking subsequences if necessary, we may assume that each of the sequences $g_{j,n}$, $n = 1,2,\cdots$, as well as each of the sequences

$$\lambda_A^{j,n} , \quad n = 1,2,\cdots , \tag{16}$$

converges. Clearly

$$0 \leq \Lambda_A^j = \lim_{n\to\infty} \lambda_A^{j,n} \leq 1 .$$

Since the sequences 16 are bounded uniformly in j,n , and A, and the series 9 is absolutely convergent, we have

$$\lim_{n\to\infty} h_{j,n} = \sum_i \Lambda^j_{A(i)} h_{A(i)} ,$$

pointwise on R. Thus

$$H = \lim_{n\to\infty} f_{1,n} \cdots f_{p,n} = \exp(-\sum_{i,j} \Lambda^j_{A(i)} h_{A(i)}) , \tag{17}$$

pointwise boundedly on R. Set

$$g = \lim_{n\to\infty} g_{1,n} \cdots g_{p,n} ,$$

$$H_n = \exp(\sum_{i=1,j=1}^{n,p} \Lambda^j_{A(i)} h_{A(i)}) ,$$

for $n = 1,2,\cdots$. By lemma 8.2.1, each H_n is a bounded l.a.m, and the convergence of the series 9 implies that

$$\lim_{n\to\infty} H_n H = 1 , \tag{18}$$

pointwise boundedly on R. If H and each H_n had a global branch, we would be done. Usually neither has, so the system of multipliers admitted by R_0 must be invoked.

From lemma 8.2.3 and equation 15, we see that $\theta_{g \circ i_0} \circ i_{\mathcal{B}} = 1$. As in the proof of lemma 1, we thus obtain from equations 5.1.2, 10, and 17,

$$\exp(+2\pi i \sum_{j=1}^{p} \Lambda^j_A) = \theta_{gH}(Z_A) = \theta_{|F|}(Z_A) = 1 ,$$

for each component A. Thus $\theta_{H_n} \circ i_{\mathcal{B}} = 1$ for each n. Define $v(n) \in \mathcal{X}(R_0)$ by

(19) $$\nu(n) \circ p_A = (\theta_{H_n} \circ p_{\mathcal{B}} \circ k^{-1})^{-1}$$

$$\nu(n) \circ i_A = 1 \quad ,$$

where the maps are those in diagram 4, and set $T_n = \varepsilon_{\nu(n)}$, $n = 1,2,\cdots$. Clearly $\theta_{T_n H_n} = 1$, so $T_n H_n$ has a global branch F_n on R . Passing to a subsequence if necessary, set $T = \lim_{n\to\infty} T_n$. By equation 18,

$$\lim_{n\to\infty} |F_n F| = \lim_{n\to\infty} T_n H_n gH = Tg \circ i_0 \ ,$$

pointwise boundedly on R . Thanks to lemma 8.2.3 and equation 12, $\theta_{Tg} \circ i_0 = 1$. Using diagram 4, in particular using the isomorphism k , we have $\theta_{Tg} \circ p_A = 1$. By construction, $\theta_{Tg} \circ i_A = 1$. Thus $\theta_{Tg} = 1$, so Tg has a global branch S on R_0 .

By taking another subsequence and changing S by a multiplicative constant of modulus 1 if necessary, we may assume that

(20) $$\lim_{n\to\infty} F_n F = S \circ i_0$$

in the β topology. Clearly

$$|S| = \lim_{n\to\infty} T_n g_{1,n} \cdots g_{p,n} \ ,$$

so an application of definition 5.2.1C to the system of multipliers 5 admitted by R_0 shows that S is exterior in $H^\infty(R_0)$. It is trivial that the restriction map $f \longrightarrow f \circ i_0$, taking $H^\infty(R_0)$ into $H^\infty(R)$, is β continuous. Thus $1 \in \mathrm{Cl}\, S \cdot H^\infty(R)$. From equation 20, we see that $\mathrm{Cl}\, S \cdot H^\infty(R)$ $\mathrm{Cl}\, F \cdot H^\infty(R)$, so F is exterior. This completes the proof of theorem 5.3.2.

8.4. This section will be devoted to the proof of theorem 5.3.3. We shall refer to $\alpha > 0$, which is either a critical value of $G(a_0,z)$ or $G_0(a_0,z)$, simply as a critical value. It is clear from the proof of lemma 6.2.2 that the set of critical values is a discrete subset of R . By contrast with the previous section, it will suffice to let the coefficient ring for homology be $\mathbb{R}$.

Proof of Theorem 5.3.3: The key to the proof is the following observation, which is a simple consequence of lemma 8.1.3 and the connectivity of the sets R^α and R_0^α: For each non-critical value $\alpha > 0$, $\beta(R^\alpha)$ is the number of critical points of $G(a_0,z)$ in R^α , and $\beta(R_0^\alpha)$ is the number of critical points of $G_0(a_0,z)$ in R_0^α . Thus the condition 5.2.4 for the G-boundedness of R becomes

(1) $$-\int_0^\infty \alpha d\beta(R^\alpha) < \infty .$$

By an integration by parts, equation 1 is equivalent to

(2) $$\int_0^\infty \beta(R^\alpha)d\alpha < \infty .$$

(The idea of integrating by parts here is borrowed from Widom [50]. In [26], we produced a very clumsy proof of the Blaschke region case of theorem 5.3.3, not using an integration by parts). The G-boundedness of R_0 implies

(3) $$\int_0^\infty \beta(R_0^\alpha)d\alpha < \infty .$$

These inequalities reduce the proof of the theorem to a careful estimate of $\beta(R^\alpha)$ in terms of $\beta(R_0^\alpha)$ and the number of compact components of $R_0 \sim R^\alpha$.

The integral 1 is a Stieltjes integral, and the functions $\beta(R^\alpha)$ and $\beta(R_0^\alpha)$ must, of course, be changed slightly in order to be right continuous. First, consider $\beta(R^\alpha)$ and $\beta(R_0^\alpha)$ restricted to the set of non-critical $\alpha > 0$. Both functions are non-increasing, have jump discontinuities only on the discrete set of critical values, and are constant between successive critical values.

Extend the restrictions of $\beta(R^\alpha)$ and $\beta(R_0^\alpha)$ to all of $(0,\infty)$ by right continuity. For simplicity, we shall not rename the two extensions, but from now on, $\beta(R^\alpha)$ is guaranteed to be the first Betti number of R^α only if α is not a critical value of $G(a_0,z)$. A similar caveat applies to $\beta(R_0^\alpha)$. Neither caveat will cause difficulty in the sequel.

Given $\alpha > 0$ not a critical value, we define an auxiliary region S^α as follows: Note that $R^\alpha \subseteq \mathrm{Int}\, R_0^\alpha$, since $G_0(a_0,z) > G(a_0,z)$ on R. Let C^α be the union of the compact components of $\mathrm{Cl}(R_0 \sim R^\alpha)$, and set $S^\alpha = R^\alpha \cup C^\alpha$. By lemma 8.1.1, each component A of $\mathrm{Cl}(R_0 \sim R^\alpha)$ has the property that $\partial A \subseteq \partial R^\alpha$. Using the covering 6.2.9, we see that S^α is a closed Jordan subregion of R_0 and $\partial S^\alpha \subseteq \partial R^\alpha$. Further, if A is a component of C^α, we have

$$\alpha = G(a_0,z) \leq G_0(a_0,z) ,$$

on ∂A. By the minimum principle,

$$\alpha < G_0(a_0,z) , \tag{4}$$

on A. In particular, $S^\alpha \subseteq \mathrm{Int}\, R_0^\alpha$.

Let $0 < \gamma < \delta$ be such that the interval $[\gamma,\delta]$ contains no critical values. As in the proof that the map 8.1.9 is an excision, C^γ is a deformation retract of C^δ, and S^δ is a deformation retract of S^γ. Hence the functions $\beta(S^\alpha)$ and $n(C^\alpha)$ are constant on intervals containing no critical values. Thus we may extend both $\beta(S^\alpha)$ and $n(C^\alpha)$ to all of $(0,\infty)$ by right continuity. For simplicity, we shall not rename the extensions. We shall divide the rest of the proof into a series of lemmas:

<u>Lemma</u> 1: $\beta(R^\alpha) \leq \beta(S^\alpha) + n(C^\alpha)$.

<u>Proof</u>: It suffices to consider non-critical $\alpha > 0$. Given such an α, consider the commutative diagram

(5)
$$\begin{array}{ccccccc} H_1(\partial C^\alpha) & \xrightarrow{i_1} & H_1(R^\alpha) & \xrightarrow{j_1} & H_1(R^\alpha,\partial C^\alpha) & \longrightarrow & H_0(\partial C^\alpha) \\ \downarrow{\scriptstyle i} & & \downarrow & & \downarrow{\scriptstyle k} & & \downarrow{\scriptstyle i} \\ H_1(C^\alpha) & \xrightarrow{i_2} & H_1(S^\alpha) & \xrightarrow{j_2} & H_1(S^\alpha,C^\alpha) & \longrightarrow & H_0(C^\alpha) \end{array}$$

induced by the inclusion map between space pairs,

$$p: \ (R^\alpha,\partial C^\alpha) \longrightarrow (S^\alpha,C^\alpha) \ .$$

Thanks to the excision 8.1.9, p is an excision. Let $G_1 = j_1(H_1(R^\alpha))$ and $G_2 = j_2(H_1(S^\alpha))$. For simplicity, denote the restriction of k to G_1 by k also. Because the rows of diagram 5 are exact and the map i is an isomorphism in dimension 0 , the map

$$k: \ G_1 \longrightarrow G_2 \ ,$$

is onto, and thus is an isomorphism.

Because there exist non-compact components of $Cl(R_0 \sim R^\alpha)$, there exist one or more contours of ∂R^α not intersecting ∂C^α . As in the derivation equation 8.1.10,

$$H_2(R^\alpha,\partial C^\alpha) = H_2(S^\alpha,C^\alpha) = 0 \ .$$

Thanks to the homology exact sequence, i_1 and i_2 are injections. Thus we have the commutative diagram with exact rows,

(6)
$$\begin{array}{ccccccccc} 0 & \longrightarrow & H_1(\partial C^\alpha) & \longrightarrow & H_1(R^\alpha) & \longrightarrow & G_1 & \longrightarrow & 0 \\ & & \downarrow{\scriptstyle i} & & \downarrow & & \downarrow{\scriptstyle k} & & \\ 0 & \longrightarrow & H_1(C^\alpha) & \longrightarrow & H_1(S^\alpha) & \longrightarrow & G_2 & \longrightarrow & 0 \end{array} \ .$$

Because k is an isomorphism, we already know that $\dim G_1 = \dim G_2$. Since ∂C^α is a disjoint union of circles, $\dim H_1(\partial C^\alpha) = n(\partial C^\alpha)$. If A is a component of C^α with r boundary contours, then $\dim H_1(A) \geq r - 1$ by the classification theorem for surfaces or by a simple direct calculation. Thus, from diagram 6, we obtain,

$$\begin{aligned}\dim H_1(R^\alpha) &= \dim H_1(\partial C^\alpha) + \dim(G_1) \\ &= n(C^\alpha) + (n(\partial C^\alpha) - n(C^\alpha)) + \dim(G_1) \\ &\leq n(C^\alpha) + \dim H_1(C^\alpha) + \dim G_2 \\ &= n(C^\alpha) + \dim H_1(S^\alpha) .\end{aligned}$$

This establishes lemma 1.

<u>Lemma</u> 2: $\beta(S^\alpha) \leq \beta(R_0^\alpha)$.

<u>Proof</u>: Again it suffices to consider non-critical $\alpha > 0$. Given such an α , consider the portion of the exact homology sequence

$$H_2(R_0^\alpha, S^\alpha) \longrightarrow H_1(S^\alpha) \xrightarrow{\ i\ } H_1(R_0^\alpha) . \tag{7}$$

As in the proof of lemma 8.1.2, each component of $R_0^\alpha \sim \operatorname{Int} S^\alpha$ contains a boundary contour not intersecting ∂S^α . Also, each component is a compact bordered Riemann surface. From equation 8.1.10 and the excision 8.1.9, we see that

$$H_2(R_0, S^\alpha) = H_2(R_0^\alpha \sim \operatorname{Int} S^\alpha, \partial S^\alpha) = 0 .$$

Consequently, the map i in diagram 7 is an injection, so $\dim H_1(S^\alpha) \leq \dim H_1(R_0^\alpha)$.

Let A_i , $i = 1,2,\cdots$, be a univalent enumeration of the components of $R_0 \sim R$, and let $a_i \in A_i$, $i = 1,2,\cdots$, be a sequence such that

(8) $$\sum_{i=1}^{\infty} G_0(a_0,a_i) < \infty .$$

The existence of such a sequence is guaranteed by definition 5.3.2A.

Lemma 3: $\int_0^{\infty} n(C^{\alpha})d\alpha \le \sum_{i=1}^{\infty} G_0(a_0,a_i)$.

Proof: The key to the proof is that each component of C^{α} contains at least one component of $R_0 \sim R$ by the minimum principle applied to $G(a_0,z)$. Let $k_1 < k_2$ be positive non-critical values such that the interval $[k_2,\infty)$ contains no critical values. Let $\alpha(1) < \alpha(2) < \cdots < \alpha(n)$ be the critical values in $[k_1,k_2]$. Of course, the jumps of $n(C^{\alpha})$ occur at the $\alpha(i)$'s. Choose $\varepsilon > 0$ so small that $\alpha(i) + \varepsilon < \alpha(i+1) - \varepsilon$ for $i = 0,1,\cdots n$. Here $\alpha(0) = k_1$ and $\alpha(n+1) = k_2$. Inductively choose finite sequences

$$a_A , \quad A \text{ is a component of } C^{\alpha} ,$$

for each $\alpha = \alpha(i) - \varepsilon$ or $\alpha(i) + \varepsilon$, where $i = 1,2,\cdots n$, as follows: For each component A of C^{α} , $a_A \in (R_0 \sim R) \cap A$, and $a_A \in \{a_j : j = 1,2,\cdots\}$. If $B \subseteq A$ is the component of $C^{\alpha(i)-\varepsilon}$ (respectively $C^{\alpha(i)+\varepsilon}$) containing a_A , where A is a component of $C^{\alpha(i)+\varepsilon}$ (respectively $C^{\alpha(i+1)-\varepsilon}$) , then $a_B = a_A$. In this induction, we start at $i = n + 1$ and count backwards to 0 .

From inequality 4, we see that

(8) $$\alpha(i)(n(C^{\alpha(i)-\varepsilon}) - n(C^{\alpha(i)+\varepsilon})) \le \sum G_0(a_0,a_A) , \quad i = 1,2,\cdots n ,$$

where the sum is taken over all components A of $C^{\alpha(i)-\varepsilon}$ such that $a_A \neq a_B$ for any component B of $C^{\alpha(i)+\varepsilon}$. To obtain inequality 8, we have used the fact that each component of $C^{\alpha(i+1)-\varepsilon}$ contains one and only one component of $C^{\alpha(i)+\varepsilon}$, and vice-versa. From inequality 8, we easily obtain

$$-\int_{k_1}^{k_2} \alpha dn(C^{\alpha}) \le \sum_{i=1}^{\infty} G_0(a_0,a_i) .$$

An integration by parts establishes the lemma.

Theorem 5.3.3 follows easily from inequalities 2 and 3 and these three lemmas.

REFERENCES

1. L. Ahlfors and L. Sario, Riemann Surfaces, Princeton, Princeton, N. J., 1960.

2. E. Artin and H. Braun, Introduction to Algebraic Topology, Merrill, Columbus, Ohio, 1969.

3. A. Beurling, On Two Problems Concerning Linear Transformations in Hilbert Space, Acta. Math. 81 (1949), 239-255.

4. R. Buck, Algebraic Properties of Classes of Analytic Functions, Seminars on Analytic Functions, vol. 2, Princeton University Press, Princeton, N. J., 1957, 175-188.

5. C. Constantinescu and A. Cornea, Ideale Ränder Riemannscher Flächen, Ergebnisse der Mathematik und Ihrer Grenzgebiete 32, Springer, Berlin, 1963.

6. M. Day, Normed Linear Spaces, Ergebnisse der Mathematik und Ihrer Grenzgebiete 21, Springer, Berlin, 1958.

7. S. Fisher, Rational Functions, H^∞ and H^p on Infinitely Connected Domains, Illinois J. Math. 12 (1968), 513-523.

8. S. Fisher, Bounded Approximation by Rational Functions, Pacific J. Math. 28 (1969), 319-326.

9. F. Forelli, Bounded Holomorphic Functions and Projections, Illinois J. Math. 10 (1966), 367-380.

10. T. Gamelin, Uniform Algebras, Prentice Hall, Englewood Cliffs, N. J., 1969.

11. M. Hasumi, Invariant Subspace Theorems for Finite Riemann Surfaces, Canad. J. Math. 18 (1966), 240-255.

12. M. Hasumi and T. Srinivasan, Doubly Invariant Subspaces II, Pacific J. Math. 14 (1964), 525-535.

13. K. Hayashi, Une Frontière des Surfaces de Riemann Ouvertes et Applications Conformes, Kodai Math. Sem. Rep. 14 (1962), 169-188.

14. M. Heins, On the Lindelöf Principle, Ann. of Math. (2) 61 (1955), 440-473.

15. M. Heins, Lindelöfian Maps, Ann of Math. (2) 62 (1955), 418-446.

16. M. Heins, Selected Topics in the Classical Theory of Functions of a Complex Variable, Holt, Rinehart and Winston, New York, 1962.

17. M. Heins, Hardy Classes on Riemann Surfaces, Lecture Notes in Math. 98, Springer, Berlin, 1969.

18. H. Helson, Lectures on Invariant Subspaces, Academic Press, New York, 1964.

19. E. Hille, Analytic Function Theory, Vol. I, Ginn, Boston, 1959.

20. E. Hille, Analytic Function Theory, Vol. II, Ginn, Boston, 1962.

21. E. Hille and R. Phillips, Functional Analysis on Semi-groups, 2nd ed., American Mathematical Society Colloquium Publications, Vol. XXXI, American Mathematical Society, Providence, R. I., 1957.

22. K. Hoffman, Banach Spaces of Analytic Functions, Prentice Hall, Englewood Cliffs, N. J., 1962.

23. W. Hurewicz and H. Wallman, Dimension Theory, Princeton, Princeton, N. J., 1948.

24. C. Kennel, Locally Outer Functions (to appear).

25. R. Nevanlinna, Uniformisierung, Die Grundlehren der Mathematischen Wissenschaften LXIV, Springer, Berlin, 1953.

26. C. Neville, Ideals and Submodules of Analytic Functions on Infinitely Connected Plane Domains, Thesis, University of Illinois at Urbana-Champaign, 1972.

27. C. Neville, Invariant Subspaces of Hardy Classes on Infinitely Connected Plane Domains, Bull. Amer. Math. Soc. 78 (1972), 857-860.

28. M. Newman, Elements of the Topology of Plane Sets of Points, 2nd ed., Cambridge, Cambridge, 1954.

29. M. Parreau, Sur les Moyennes des Fonctions Harmoniques et Analytiques et la Classification des Surfaces de Riemann, Ann. Inst. Fourier (Grenoble) 3 (1951), 103-197.

30. M. Parreau, Fonction Charactéristique d'une Application Conforme, Ann. Fac. Sci. Univ. Toulouse 19 (1955), 175-189.

31. A. Pfluger, Theorie der Riemannschen Flächen, Die Grundlehren der Mathematischen Wissenschaften LXXXIX, Springer, Berlin, 1957.

32. A. Read, A Converse of Cauchy's Theorem and Applications to Extremal Problems, Acta. Math. 100 (1958), 1-22.

33. H. Royden, Boundary Values of Analytic and Harmonic Functions, Math. Z. 78 (1962), 1-24.

34. L. Rubel, Bounded Convergence of Analytic Functions, Bull. Amer. Math. Soc. 77 (1971), 13-24.

35. L. Rubel and A. Shields, The Space of Bounded Analytic Functions on a Region, Ann. Inst. Fourier (Grenoble) 16 (1966), 235-277.

36. L. Rubel and A. Shields, The Failure of Interior - Exterior Factorization in the Polydisk and the Ball, Tôhoku Math. J. 24 (1972), 409-413.

37. L. Ruble and J. Ryff, The Bounded Weak-star Topology and the Bounded Analytic Functions, J. Funct. Anal. 5 (1970), 167-183.

38. W. Rudin, Analytic Functions of Class H^p, Trans. Amer. Math. Soc. 78 (1955), 46-66.

39. W. Rudin, Essential Boundary Points, Bull. Amer. Math. Soc. 70 (1964), 321-324.

40. D. Sarason, The H^p Spaces of an Annulus, Memoirs of the American Mathematical Society 56, American Mathematical Society, Providence, R. I., 1965.

41. H. Schaefer, Topological Vector Spaces, Macmillan, New York, 1966.

42. M. Spivak, Calculus on Manifolds, Benjamin, New York, 1965.

43. G. Springer, Introduction to Riemann Surfaces, Addison - Wesley, Reading, Mass., 1957.

44. M. Tsuji, Potential Theory in Modern Function Theory, Maruzen, Tokyo, 1959.

45. M. Voichick, Ideals and Invariant Subspaces of Analytic Functions, Trans. Amer. Math. Soc. 111 (1964), 493-512.

46. M. Voichick, Extreme Points of Bounded Analytic Functions on Infinitely Connected Regions, Proc. Amer. Math. Soc. 17 (1966), 1366-1369.

47. M. Voichick, Invariant Subspaces on Riemann Surfaces, Canad. J. Math. 18 (1966), 399-403.

48. H. Weyl, Die Idee der Riemannschen Fläche, Teubner, Leipzig, 1913.

49. H. Weyl, Die Idee der Riemannschen Fläche, 3rd ed., Teubner, Stuttgart, 1955.

50. H. Widom, The Maximum Principle for Mulpiple - valued Analytic Functions, Acta. Math. 126 (1971), 63-82.

51. H. Widom, Hp Sections of Vector Bundles over Riemann Surfaces, Ann. of Math 94 (1971), 304-324.

University of New Hampshire
Durham, New Hampshire 03824

Current Address:
Central Connecticut State College
New Britain, Connecticut 06050